Chronicles of a Farmer's Life

The Incredible Rise of Bennedicter Mhlongo

With Duncan Mnisi & Faith Ngwenya

Chronicles of a Farmer's Life By Bennedicter Mhlongo
As told to Duncan Mnisi and Faith Ngwenya

First Published in August 2025
By Benica Projects Pty LTD
Rooiboklaagte
Acornhoek
1360

Typesetting: Duncan Mnisi
Cover design: Duncan Mnisi
Cover Photo: Bennedicter Mhlongo
Edited by: Goodenough Mashego
Foreword by: Labani Mgimeti
Inside photo credits: Bennedicter Mhlongo

For more information and enquiries,
Email: info@benicafarm.co.za

Dedication

This book is dedicated to my family, to all small-scale farmers across the continent, to everyone who believed in me from day one, and to the Creator of heaven and earth-God.

Chapters

Foreword

'Sesi wa mehe', that's how, Bennedicter, who we call Benica greets me whenever we catch up on the phone. '*Sesi wa mehe*' is an affectionate Xitsonga greeting meaning 'My sister'.

Benica embodies the Biblical phrase, 'And the Word became flesh'. What you will read from this book is not a fictional rags to riches story, but testimony of a life lived and still being lived to the fullest.

Benica Mhlongo: Chronicle of a Farmer Life is not a book exclusively for farmers, but rather everyone who feels they are about to give up; or someone who has thrown in the towel thinking it's over. I promise you, after reading this book you will be prompted to go and pick up that towel, wipe yourself and get back on the road.

This book is about resilience, perseverance, persistence, lessons, growth, agility and adaptation. This is a go-to manual when you feel like you are stuck with nowhere else to go.

Benica was determined to succeed regardless of many roadblocks he encountered along the way. He turned his losses to lessons which in turn strengthened his resolve.

Benica Mhlongo: Chronicle of a Farmer Life gives evidence to the saying 'You are not responsible for the family you were born in, but rather the family you build'. It will help you understand that your foundation is a fundamental key to your future, to never despise your humble beginnings.

This farmer didn't allow his circumstances to create resentment, bitterness and anger, however through discipline, he forged his way.

As an orphaned child, he didn't allow 'societal stigma' to cloud his judgement. He embarked on the new journey in pursuit of the unknown. His dedication, eagerness, work ethic and commitment raised him from the dust and placed him on the table with nobles. His personal story illustrates a picture of divine alignment and purpose.

Divine intervention has always played a role in Benica's life. In 2008 Benica was appointed Personal Assistant to then Chief Executive Officer of Tintswalo Hospital in Acornhoek, Mrs M.E Zwane.

Before I proceed, please allow me to introduce myself. My name is Labani Mgimeti, nwa'Shilubana, born and bred in Shiluvane Village. I am a Tsonga Princess of the Nkuna Royal Family and the founder of Hiteka Kitchen. Hiteka Kitchen is a healthy school lunch company based in Kempton Park, Johannesburg. The school's core focus is remodelling traditional tuckshops in South Africa's schools.

The Zwane I mentioned above is MamaHulu, my mother's older sister. Benica and I meet 15 years since he worked as MmaHulu's PA. We meet as participants of a Social Entrepreneurship Programme organised by Fetola in partnership with First National Bank.

Our relationship sparked from there. When he was invited to exhibit Benica Farm at the Sandton Convention Centre, he humbly requested my support. It was such an honour and privilege to work closely with him.

We later worked together in other projects as well, including the curating of Benica Farm website. This experience has taught me the importance of relationships. We live in a society that chases money instead of cultivating meaningful relations; unaware that the relationships you build are a direct result of your choices and actions. Investing time, effort, and care into relationships is crucial for success and well-being.

Benica's roots are so deep and strong that even when the wind blows and a storm rages, he remains standing. Despite resistance and opposition, he is confident about who he is and who is in him. Such an attitude is thanks to his late father who instilled in him the importance of trusting in God.

Benica Mhlongo: Chronicle of a Farmer paints a picture of hope, humility and love in building a legacy that will last for generations to come. A picture of trusting where you are in life, removing all doubt and focusing on the present while taking life one day at a time. It's always important to focus your mind on your dreams.

This is evident in his career shift; his destiny has always been farming, the nursing season was meant to give him a view of how much society is in dire need of nourishment. Healing goes hand in hand with the soil. After all, we are from the soil and it's only what comes from the soil that can heal us.

Today, Benica is able feed multitudes. He wakes up every morning with a drive to bring about transformation in his community, be it distributing food to the needy, collaborating with others in the value chain and grooming future agropreneurs; this is the heart of Benica Farm.

In conclusion, I urge you to go ahead and read Benica's Chronicles. It will give you a clear view of your own life, your experiences and you will recognise that your life is full of seasons. Benica was intentional when writing this book, he left no stone unturned.

Benica Mhlongo: Chronicle of a Farmer should be in your basket for must-read titles. I recommend it as a manual for entrepreneurs alike.

'Every day when I wake up, my life revolves around solving people`s food problems. To be me, greatness

means inspiring others to go out and inspire even more. We are all born to be great. I believe I was born with a divine purpose: to inspire and touch lives through my work and passion', Benica The Farmer

Preface

A Dream that emerged from a backyard garden. One Step at the Time!

My name is Bennedicter Mhlongo, Benica to many. I was born on 29 July 1984. I am the first-born son of Assah and Wisani Mhlongo. On the day I was born then apartheid Prime Minister P.W. Botha and Foreign Minister Pik Botha began a tour of several European countries, including Austria, Belgium, France, Great Britain, Italy, Portugal, Switzerland, and West Germany. The aim of their visits was to strengthen diplomatic ties and potentially ease international pressure related to South Africa's apartheid policies.

I grew up in a small village of Rooiboklaagte, outside the small town of Acornhoek. Back then the village formed part of the Bantustan of Gazankulu, a wasteland for Xitsonga and Shangaan speakers created in 1973 by the then apartheid government.

Acornhoek lies adjacent to the tourism and agriculture-rich dorpie of Hoedspruit. They both dot the R40, a 310-kilometre provincial road that begins in the mining town of Phalaborwa and stretches all the way the Eswatini border town of Bolembu.

My village of Rooiboklaagte is surrounded by the beauty of nature and the warmth of community. It has an excellent view of the iconic Drakensberg Mountain, an escarpment that stretches for 1000 kilometres through many provinces. Rooiboklaagte is one of those laid-back villages where everyone knows everyone. Growing up in the village, life was simple but offered daily life lessons. Here you'd learn about the importance of hard work, humility, and valuing family.

From a young age, I knew I wanted more than the confines of village life. I dreamt of making a difference, of telling my story, and of inspiring others to chase their own dreams. There was always a voice that told me that one day I will become a successful entrepreneur.

As a child I watched my family work tirelessly on our small farm, tending to crops and livestock, finding joy in the simple acts of planting seeds and watching them grow.

These early experiences instilled in me a deep respect for the land and a desire to explore the world beyond my small village. But even as I nurtured my dreams, I remained grounded in my cultural values. Those are values that have shaped the person I am today. My journey was not without challenges. There were moments of doubt, times when I wondered if my dreams were too big for someone like me, especially living in a rural area with limited resources. However, each obstacle became a stepping stone, pushing me to work harder, to learn more, and to believe that I could achieve whatever I set my mind on. That belief was fueled by a desire not just for personal success but to uplift others from similar backgrounds.

Before I go further, let me introduce you to Duncan Mnisi, a remarkable individual whose personal journey complements and enriches the narrative. Duncan is a storyteller whose work spans journalism, literature, radio, and visionary leadership. His journey began in 2022 when he published his first book *Visionary*. The book was a revelation to his ability to tell compelling stories from the depths of human experience.

Inspired by the social impact of his first book, in 2023 Duncan published his second work titled *Truth Revealed.* It is a book that delves into the raw realities of life and encourages readers to confront uncomfortable truths. His professional path is equally inspiring. In February 2024, he joined *Timeless News* as a journalist, where he started sharpening his skills as a news reporter.

Within a short space of time he moved to Immaculate Media, continuing to produce meaningful, impactful stories. His radio journey began at R FM Media House in Acornhoek, a community-based station where he started as a news reader. The Acornhoek of Duncan and of my childhood has changed dramatically. It is no longer the one-horse dorpie I grew up in but a bourgeoning working to middle peri-urban settlement that boasts media houses and countless opportunities. Duncan's talent and charisma quickly caught the attention of people who matter, leading him to host the Mid-Day Show on RFM for six months. He later co-hosted the lucrative slot of the Drive Show. He is now the host of the Mid-Morning Chats Show. He brings insight and authenticity that keeps listeners engaged and inspired.

Duncan's cultural footprint extends beyond journalism. In 2024, he published *Ku Tiyisela Swa Pfuna*, a groundbreaking Xitsonga work that captures the rich narratives and perspectives of the people. His writings reflect a deep understanding of human experiences, and his passion for storytelling is evident in every word.

He is also a dedicated content creator on platforms like YouTube, sharing personal development insights and life lessons to motivate and uplift others. He holds a National Diploma in Information Technology (IT) from the Vaal University of Technology. Apart from his professional life, Duncan is a family man. He balances his multiple careers with his roles as a father, brother, and community member, embodying the values of love, respect, and resilience.

His journey is a shining example of how dedication and a positive mindset can transform lives.

The next person I am going to introduce you to is Faith Ngwenya. Faith's story is equally compelling and is deeply rooted in culture, education, and a quest for personal growth. She grew up on the outskirts of Manyeleti village, specifically in Seville C. Under apartheid, Manyeleti was famous for being a section of the vast Kruger National Park that was open and accessible to Back people from Gazankulu's Mhala and Lebowa's Mapulaneng regions. Villages adjacent to the reserve to this day have both Amashangana and Mapulana communities.

Growing up in culturally-rich Seville C shaped Faith's understanding of tradition and community. She attended Manyeleti Primary School before moving on to finish her matric at Dayimani High School.

From an early age, Faith was captivated by literature. She loved to read, write, debate, and imagine worlds beyond her rural surroundings. Her fascination with history and human rights grew alongside her love for storytelling. These passions were more than hobbies; they became a calling. Faith's interest in advocating for others, understanding societal issues, and preserving cultural norms fueled her desire to make a difference. Although her pursuit for a law degree at the University of Venda was interrupted; her passion for human rights and storytelling remained strong.

After leaving university, Faith faced the challenge of job hunting in a competitive world. Her breakthrough came when she responded to a Facebook post from The Timbavati Foundation, seeking environmental educators. She applied and was fortunate enough to be selected. Working as an environmental educator allowed her to connect with nature and communities, and her passion for writing surfaced. Her role evolved into content creation, where she poured her heart into words, writing stories, raising awareness and advocating for her community and culture.

Faith's background is rich with tradition and values. Coming from a family rooted in cultural norms, she learnt the importance of maintaining her identity and passing it on her daughters. Writing became her voice and her way of advocating for those who cannot speak for themselves.

As a shy person Faith is reserved and quiet. It is through her writing that she finds her strength and purpose. Faith envisions herself in the next five years as a published author, sharing both fiction and non-fiction stories that inspire and educate.

Our collective stories; mine, Duncan and Faith's intertwine to create a tapestry of resilience, cultural pride, and hope. They showcase that regardless of where you start, your dreams are valid, and your voice matters. These stories highlight the importance of perseverance, community and staying true to oneself.

This book, *Benica Mhlongo: Chronicle of a Farmer* aims to not only tell my story but to inspire others to believe in their potential. It's a testament to the fact that with faith, hard work, and courage, anyone can rise above their circumstances and achieve greatness.

As you read through these pages, I invite you to see yourself in these stories; your struggles, your hopes, your victories.

Let them remind you that you are capable of incredible things, and that your journey is just beginning. Whether you come from a small village or a bustling city, your dreams are valid, and your story deserves to be told. Together, with Duncan's masterful storytelling and Faith's authentic voice, we hope this book sparks curiosity, ignites inspiration, and encourages you to chase your own dreams - no matter how impossible they may seem at first.

This is more than just a story of one man's rise. It's a celebration of life, culture, resilience and triumph of the human spirit. So, turn the pages with an open heart and mind, and prepare to embark on a journey that could change your life.

Introduction

"The ultimate goal of farming is not the growing of crops, but the cultivation and perfection of human beings." – Masanobu Fukuoka

No man is truly an island. To claim that I am a self-made farmer, a leader, or a visionary born solely from egoism or self-centered ambition would be both arrogant and ungrateful. Truth is, my journey was shaped by countless hands, hearts and the minds of people from all walks of life who played a pivotal role in my success. From my departed parents, Wisani and Assah Mhlongo, my siblings, mentors, community members, friends and even strangers who believed in me when I doubted myself; every influence, every lesson learned, has contributed to the person I am today.

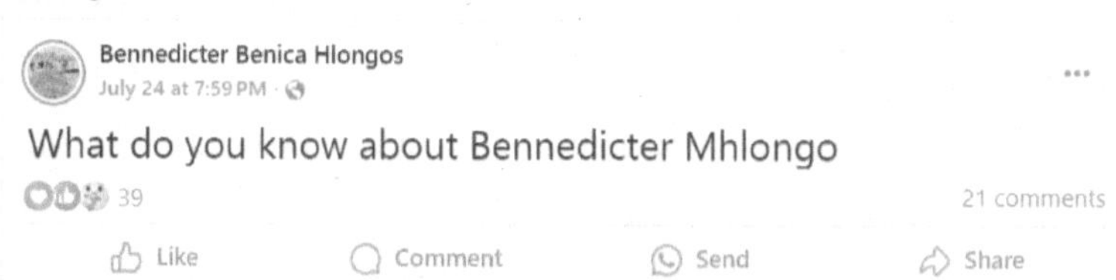

Before embarking on the telling of this story, I couldn't help but wonder how people perceived me. To test the waters, I took to social media and asked a:

'What do you know about Bennedicter Benica Mhlongo?'

The responses I received were surprising, funny, and eye-opening. Here's what some people said;

'Benica is a self-driven businessman, devoted to family and business, and very intentional about what he wants. He's a 100% pure hustler, from the streets to big retail stores.' - *Matome Potatoe Farmer, Limpopo*

'He's highly driven and a leader in his space. He's a pioneer, a doer, and visionary. He's a food producer and job creator in Mpumalanga, making agriculture fashionable.' - *Lebo Mashigo commercial Chicken Farmer, Mpumalanga*

'I know him as the master of food security, contributing to the production of veggies and fighting unemployment in his community. He plays a big role in skills development through his farm, Benica Farm.' - *Templeton Tshepiso Jantjies, Commercial farmer, North West*

'He's the CEO at Benica Farm, producing quality, kind, down-to-earth, and always sharing knowledge. To us young farmers, he's our Idol.' - *Kwazikwakhe Kunene Small Scale Farmer, KZN*

'A mentor of note.' - *David Mnisi, Regional Manager NYDA*

'Benica is not only passionate about farming but also a community builder, a strong and selfless person. I love how you encourage others to win and become better.'

- Alice Letaba, Semi-commercial Farmer, Lesotho

These responses remained with me as I narrated my story. I hope they will also inspire you as you read through these pages. My journey has been shaped by my passion for farming, my commitment to community, and my desire to share knowledge and inspire others.

From the very core of my being, farming is more than just a livelihood, it's the heartbeat of everything I do. It's a symphony of resilience, hope, community, and personal growth.

My story isn't just about crops and harvests; it's about perseverance, innovation, and an unwavering dedication not only to the land but also to inspiring others to find their roots and grow beyond their limitations.

In this book I chronicle my journey; from humble beginnings to becoming a commercial farmer, social entrepreneur, mentor, and community leader. It reflects how farming and life are intertwined, shaping my character, aspirations, and a legacy I hope to leave behind. I deliberately chose this farming lifestyle.

In it, I find purpose and passion. We are the children of a father who invented hope—hope for a better future, hope for resilience in the face of adversity. Even though many people dismiss agriculture as a flash in the pan or a joke, I see it as the foundation of our existence. I firmly believe that the next generation of billionaires will come from the agricultural sector.

The above is not just a statement but a fact of life. Believe me when I say that agriculture is the heartbeat of this universe; and us farmers hold the power to create a better world. Through opportunities, innovation, and kindness, we can build a legacy that whispers hope into the hearts of our children and communities. A world where no one is ashamed to live, where dignity and sustainability are at the core.

I am among a fortunate few who have chosen to make this world a better place through farming. And I want you, whether you are a farmer at heart or a city dweller craving a glimpse into rural truth, to walk in the muddy boots of those who feed the world. As former Burkina Faso revolutionary leader Thomas Sankara famously opined, 'He who feeds you controls you'. We are not here to claim our power in society; rather, to send a message, to challenge perceptions and to remind society that farmers are vital, powerful, and deserving of respect.

Despite what many say that farming is a white man's vocation, I am here to turn that narrative on its head. Farming is every person's responsibility. It doesn't matter your race, your background, or where you come from. At the end of the day, we all eat. We all need provision. It's a universal truth. This book chronicles my seemingly ordinary life; a life filled with hard work, setbacks, and triumphs. I'm a living testament to that undeniable truth.

To all aspiring farmers out there; you don't need to have everything to start. You just need to know why you want to farm; and once you begin, always remember your 'why'. That's the message of hope that has carried me through for over 13 years in this industry.

Believe me when I say, every seasoned farmer was once a beginner; a nobody. That you are holding this book, that you are curious about my journey, is already a good starting point. To everyone who has been inspired by my story; I extend my deepest gratitude.

I hope my life story answers your questions, reveals the truths you've heard about me, and takes you inside my life the way you've envisioned it. If you own this book, or it was handed to you by someone who believes in its message, you've made a wise investment of both your time and resources. As you read, I invite you to explore themes of hope, resilience, leadership, and the transformative power of farming; elements that can redefine lives and communities.

The question remains: where and how did it all begin? What is the story behind my success? How did the business, Benica Farm come about?

1

Roots & beginnings

'Support your local farmer. You can't eat oil',

Unknown

Let me take this moment to reintroduce myself. As mentioned in the Preface I was born in 1984 and raised in Rooiboklaagte B, a village in the small town of Acornhoek. Rooiboklaagte B is a humble community nestled within the rich, rolling landscapes of Bushbuckridge. My childhood was a tapestry woven with both hardship and hope. My family consisted of seven members, my father, Wisani Mhlongo, who was a former mine worker and salesperson; my mother, Assah Mhlongo, who was a loving and resilient woman, and my four siblings, Adolf, Rodney, Xitshembiso and Ashante (deceased). My parents and Ashante have since passed on.

I started school at Senianya Primary School in 1991, a year after Nelson Mandela walked out of prison and began his long walk to freedom. It was a euphoric time of unbridled optimism and hope for the future. Our days in school were filled with innocence and dreams, even amid the struggles that marked our lives. Apartheid was on its last legs, and just as they say about the last kick of a dying horse, it was tough for most people in my village and school.

Before he was retrenched in 1991, my father worked as a mineworker at Dundee in KwaZulu-Natal. I remember that before he was retrenched in the 1990s our family lived a good life. Both my parents worked hard to provide us with a decent upbringing. I remember the joy of simple pleasures; playing with friends, helping my mother with house chores, and dreaming big despite our modest circumstances.

My father's entrepreneurial spirit was evident early on. After losing his job, he started and operated several small businesses, reselling locally fashionable clothes he bought in Durban.

His resilience was a lesson I carry with me to this day.

Life took a different turn when he got another job at Pilgrim's Rest Mine and later in Witbank. He later got retrenched again. It was during those times that our family faced many hardships. We started living from hand to mouth, struggling to make ends meet. At that time I was doing Grade 10 at Puledi High School.

To add insult to injury, amid the turmoil in our family, the same year my mother fell seriously ill and passed away before I could write my matric examinations in 2001. My loss, though painful, didn't affect my schooling as I later passed my Grade 12. That was a turning point in my life; a moment of profound loss and responsibility.

After my mother's passing, our family was under the custodianship of my paternal grandmother, Nwa'Sikhemisi Lilly Mhlongo and our uncles, Obed Mokoena (deceased) and Fanuel Mokoena who both played a key role in our childhood. As a first born, I suddenly found myself stepping into a role I never anticipated; I became the caregiver, the household manager, and the emotional anchor for my siblings when after my mother's funeral, my father once again left to look for a job in KZN.

Taking on adult responsibilities was no easy feat. I had to juggle school, household chores, and caring for my siblings who looked up to me as their leader and protector. I remember a time when me and a few friends formed a soccer team which's name I have since forgotten. I was not just their captain but their role model as well, someone they admired and respected. I had moved our amateur team closer to my house so I could keep an eye on them while attending to my parenting duties.

Forced to grow up fast, I learned early that leadership is not about titles but serving others and standing firm in the face of adversity.

At this point, Nwa'Sikhemisi's home was our refuge; a place where we found one meal a day, often just enough to keep us going. I vividly remember one day during school break, when I bunked to visit her, hoping to get that one pawpaw fruit I'd been craving. I was disappointed to find that one of my uncle's had already eaten it. That moment was a harsh reality check; a reminder that life was tough.

It was during 2001, during my matric year that I received the news that my father had died in a car accident. It was a family relative who informed my elders. Unfortunately, his body was never found. Rumors swirled that he might have been living in Springs or somewhere else before traveling to KZN, but the truth remained a mystery.

To this day my father's body has not been recovered. Some of our relatives went to search for his body in mortuaries they suspected he would've been taken to no avail. We continued with funeral arrangements even in the absence of the body. Despite these tragedies, I refused to let despair define me. I made a firm decision; my future would not be dictated by misfortune. I was determined to succeed.

After high school, I moved to Katlehong in the East Rand of Johannesburg to stay with uncle, Fanuel, who was a Metro Police officer in Gauteng. I was hoping to find work, but I was too young. At the time I was only

17-years old and inexperienced. Life was relentless and the struggle continued.

I however didn't give up. In 2003, through my now deceased uncle Obed I got a chance to join the Group of 30 candidates on a Limpopo government program called Poverty Alleviation, which's aim was to empower orphans and rural youth. The government, under MEC for Health Seaparo Sekwati was offering great opportunities to orphans in our area and I was identified to be part of this initiative. I was going to be trained as a nurse. I joined the group. I recall that when my younger brother Adolph met me he said, `Buti, Malumi uri uya xikolweni xa nursing. Paka tibheke vatata vata ku teka', *meaning, 'My brother, uncle says you going to a nursing school. Pack your bags they will come fetch you'.*

I froze for a moment, shocked and in disbelief.

I trained at Sekororo Nursing College. That opportunity changed my life; taking me out of despair and into a world of public service and hope.

I graduated in 2005 and was assigned to Sekororo Hospital, which was far from home in Rooiboklaagte. The following year, I transferred to Tintswalo Hospital in Acornhoek, near home. Balancing work and family responsibilities was tough, especially as I continued to support my siblings' education. Adolf was already studying Information Technology, Rodney was in healthcare, and I was juggling multiple roles. My dedication caught the attention of hospital staff, and I was recognized for my versatility and leadership.

In 2006, I was circulated in three hospital wards, surgical, maternity and TB. It was in this versatility that demonstrated my capacity to adapt and excel. My work ethic and commitment did not go unnoticed. I was soon appointed as a Personal Assistant to the hospital's Matron, Ms. M.B Moyimane. This was a turning point in my life. The new role exposed me to administration, graphic design, and communication tasks, opening my eyes to new possibilities beyond nursing.

Moyimane believed in my potential. She saw qualities in me that many overlooked. Under her mentorship, I learnt to navigate complex administrative systems, develop leadership skills, and harness my talents. I was still a nurse, but I was also transforming into a versatile professional; one ready to take on challenges and seize opportunities.

In 2008, I received a job offer from a private company, but the hospital management refused to release me, stating that I was now a valuable asset to the institution. Instead, they counter-offered by appointing me as the Personal Assistant to the Chief Executive Officer. I was moved from the matron's office to the office of the CEO, Mrs M.E Zwane.

This transition marked a major turning point in my career. Working closely with the CEO exposed me to high-level hospital operations, including governance, institutional management, and strategic engagement with the hospital board, stakeholders and provincial leadership.

I frequently attended and managed high-profile meetings, which taught me the importance of precision, confidentiality and professionalism. That office played a significant role in shaping who I am today. It was there that learnt the value of commitment, working long hours, sometimes even overnight to prepare documentation and support leadership.

Later, I was seconded to act in a senior role as Assistant Director: Communications. My key responsibilities included corporate communications, media relations, and hospital administration. I continued to study, earn certifications and build my reputation.

I decided to pursue a Diploma in Public Relations at Mapulaneng TVET College as a part-time student. With every module I took my interest in communication grew stronger.

I wanted to expand my skills, to become more than just a hospital worker. I wanted to influence, to lead and to inspire. I balanced studies with my work, often working late into the night and over weekends, to master new skills. My dedication paid off. I became the Chairperson of the Bushbuckridge Communicators Forum, representing hospital communication officers across the district, a role that further boosted my confidence and leadership qualities.

This potential was recognized when I started writing newsletters for the hospital. I was then later transferred to the Communications Office.

Throughout this journey, I was often confronted by jealousy, envy, suspicion, and workplace toxicity. Some colleagues resented my rapid rise, questioning my qualifications and accusing me of favoritism. But I proved myself through relentless effort; working extra hours, contributing beyond expectations, and refusing to be sidelined. I knew my worth, and I was determined to grow.

My potential had many of my colleagues shook. It caused many of them chest pains but I remained calm. I remember days some would give me jealous looks. What caused many a chest pains was that when they were supposed to submit a report, it had to pass through me.

They held small gossip meetings about me. Sometimes when I stepped through the hospital gates, I could see the tension that had them wondering why I had this privilege of climbing this ranks that takes most of them years to reach.

Year after year I climbed higher and higher. That attracted a lot of enemies.

Eventually, I was promoted to senior positions, taking on responsibilities in communications, media, and hospital administration. I continued to study, earn Certificates and build my reputation. Yet, amid all this professional growth, an entrepreneurial spirit was quietly awakening within me - a desire to create something of my own outside the hospital walls.

After serving eight years in healthcare, I heard a calling. Farming was calling me. I then started a backyard garden. Neighbors started buying from my garden. It was no longer just a passion; it became a purpose. My journey in healthcare, with all its ups and downs, had prepared me for this next chapter; one rooted in the land, in community, and in hope.

This is how my story began. A story of resilience, dreams, and the unyielding belief that with faith and hard work, anyone can rise from the humblest of beginnings. As we continue, I invite you to walk with me through every step of this incredible journey; through the muddy fields, the sleepless nights, the victories, and the lessons learned along the way.

What people learn:

Our origins and early experiences shape our character. Hard work, humility, and resilience are born from humble beginnings. The chapter teaches that embracing your background and learning from hardship can fuel future success.

2

Personal life & core values

'The soil is the great connector of lives, the source and destination of all' – Wendell Berry

Beyond my professional pursuits and journey and entrepreneurial ventures, I am at the core of my being a family man. My greatest pride and source of strength is my wife, Veronica Mhlongo. Together we have four wonderful children who inspire us greatly; Mohau, Wisani, Benica, and Nsuku. They are my anchor, my reason for waking up each day with purpose and humility. My family grounds me in values that transcend wealth or success; they remind me of what truly matters; love, humility, service, and the responsibility to nurture those around me. American actor Michael J Fox once said, 'Family is not an important thing, it's everything'.

Growing up, I was deeply influenced by my background as a nurse. That profession taught me the importance of caring, empathy, and community well-being.

These qualities have carried over into my approach to farming and life.

I believe that at the core of leadership lies service—serving your family, your community, and the land you till. It's this philosophy that has fueled my journey from humble beginnings to where I find myself today.

I can confidently say I am a devoted family man—principled, humble, and grounded in my beliefs. The late former US First Lady Jacqueline Kennedy Onassis once told NBC Correspondent Sander Vanocur in 1960, 'If you bungle raising your children, I don't think whatever else you do well matters very much'.

Man of God

I believe my faith as well has played a pivotal role in shaping my character and guiding my decisions. From a very young age, I found myself deeply rooted in the International Pentecost Holiness Church (IPHC). I love church, not just as a ritual but because of the love of God that sustains me. I have always believed there is a powerful unseen force, something divine that controls not just our physical lives but our destinies beyond this world.

My spiritual journey began in 1998 when my late father introduced me to the church. His unwavering faith and non-negotiable beliefs left a lasting impression. I was fourteen years old, impressionable, and eager to learn. His discipline and unshaken trust in God became the foundation of my spiritual life. I grew up understanding that discipline and morality go hand in hand. It's a simple truth that discipline shapes character and life itself. Many who walk that path will testify that discipline has the power to transform lives, instill resilience, and foster integrity.

These values; discipline, humility, integrity, are not just words to me - they are principles I strive to live by every day. They influence how I raise my children, how I interact with my community, and how I manage my land and businesses. I believe that true strength and success come from inner conviction, deeply rooted faith and solid morals.

From a religious perspective, I am convinced I had an epiphany not through church attendance but within myself. I hope you will find your own inner strength and faith, regardless of your religious or spiritual beliefs. Faith, to me, is a personal compass; something that guides me through life's storms and uncertainties. I have often thought about the unseen forces that influence our lives, and I believe that aligning oneself with divine purpose brings clarity, strength, and resilience.

In a conversation with my co-writer, Faith Ngwenya, former CEO of Tintswalo Hospital Ms. Zwane remarked that I possess a high level of spiritual caliber. She said, 'He's highly spiritual. That's my honest opinion. I have no evidence of his spiritual life, but I can feel it'. Many people have wondered about the source of my strength; how I carry myself through challenges, setbacks, and losses. I believe that my spiritual alignment, being connected to my purpose and divine guidance has been fundamental.

Usually, people imbued with spiritual talents are often burdened with a higher calling - they are intuitive, gifted, and resilient.

Letty Mhlongo

Let me share with you a chapter from my past, another story that shaped much of who I am today. It's about my life with my late wife, Letty Mhlongo, and the journey we traveled together - a journey filled with love, challenges, and lessons.

Letty and I brought to this world two beautiful children, Mohau, our firstborn, and Wisani, who is named after my late father.

Our marriage was a beautiful chapter of my life. We married in church, surrounded by family, hope, and faith. Like many young couples, we faced our ups and downs, but through it all, we built a life many dream of, stable, loving, and full of potential.

However, just a few years after beginning to build our life together, tragedy struck. At the time I was running small businesses; a couple of ventures I poured my heart into. One of those was a fish and chips operation in Tsakani, which we started in a modest six-meter container next to Dynamos Sports Ground. It was a small start, but it flourished due to our devout dedication and hard work.

Our business was thriving, and with it came jealousy. Every community has those members who are enemies of progress. It started with one neighbor claiming that our operation was a magnet for thieves; that criminals hide in the container and later rob people at night. At first, I dismissed her accusations, confident that our success was legitimate. But tensions grew, and disagreements escalated. I was stubborn, perhaps too stubborn; and insisted that we keep our business going. We eventually took the matter to the local Induna. As a local traditional authority, it is his responsibility to peacefully resolve community disputes.

The conflict left indelible scars.

During conflict resolution, one old woman`s words were, "Uta swi vona", a Xitsonga phrase meaning "You will see, and you will regret". Those words haunted me. I didn't realize at the time how deeply those words would mark my life.

Shortly after that dispute, my wife Letty fell gravely ill.

She experienced persistent headaches for nearly a month. We sought treatment at Tintswalo Hospital, then at Rob Ferreira in Nelspruit, but doctors couldn't diagnose her illness. Her health rapidly deteriorated.

During those difficult days, she openly expressed her thoughts. She sensed her time was near. She started to talk about the need for another woman in our lives, a support system for her and our children. She asked me to marry a second wife, to help care for the family, especially since Wisani was still very young.

Marrying a second wife was a foreign concept to me; something I never envisioned at my age. But my late wife's reasons were genuine and compassionate. She believed that a second wife would help carry the burden of caring for our family. She managed to convince me. The BIG question was, who will be the second wife? We looked around and eventually agreed that we were going to marry Veronica. She is a young, down-to-earth and respectful lady, also from church. Veronica comes from a polygamous family. Letty's trust and love for Veronica was evident. I respected her wishes. She had already seen Veronica as a suitable partner in her heart, even before I did.

At the time Veronica was a young woman working at KFC. She was vibrant, charismatic and grounded. Those were qualities that drew me close to her. When

Letty suggested we bring Veronica into our lives, I hesitated. Veronica's schedule was demanding, and I wondered if it would work. But my wife's conviction was strong, and she arranged for us to meet Veronica at the mall during her lunch time.

The meeting was unexpected. My wife took the initiative to introduce us at the mall; an ordinary place turned into a pivotal moment in my life. Veronica agreed to join us, and within a short space of time, arrangements were made through church elders.

As time went on, Letty's health continued to deteriorate. She knew her time was limited. She started saying her goodbyes, her last words, her final embraces. Veronica stepped into the void with grace, caring for our children, supporting me and becoming a vital part of our lives. The bond between Veronica and my children grew strong; they affectionately called her Mama Vero, recognizing her as a caring, maternal figure.

Meanwhile, Letty's health worsened. Despite our efforts, her condition deteriorated. She was admitted to Tintswalo Hospital again, and her prognosis was grim. She confided in me that she believed she wouldn't walk out. I remember that day vividly.

I was busy with my daily routine when I received a call from the hospital; an urgent request to come. When

I arrived, I finally accepted the inevitable; Letty was leaving us.

Her passing in 2014 was a devastating blow. Losing someone so close; my wife, my partner in life, the mother of my children, left a void that words cannot describe.

I have endured profound loss; parents, siblings, friends, and loved ones – each leaving behind memories etched into my soul. I often wonder what life would have been like if they were still here to witness my journey. I believe they look down from above and see the fruits of my labor and the legacy I am building.

A year after Letty's passing, I was approached by a couple of ladies from my church, expressing their interest in me – hoping I would marry again. But I knew what I wanted. I was already committed to Veronica, to building a new life together.

Two years after Letty's passing, I officially married Veronica. She was still working at KFC, often late into the night, commuting by minibus. One evening, after her shift, she nearly fell victim to robbers just outside our home, a stark reminder of the risks faced by hardworking women in our community.

That event prompted me to act. I used my savings to buy her a car, ensuring her safety on the road. But I knew that wasn't enough. Her late shifts still posed dangers. So, I proposed she leave her job and work with me at my farm. She agreed, and together, we set out on a new chapter; one that would eventually lead us to the creation of Benica Farm.

Throughout these trials, personal, professional, and spiritual, I realized that healing was essential. It was time for a change, a renewal of purpose. As I reflect on my journey, I understand that life's hardships are part of a divine plan. The pain, the losses, and the victories, they all serve a higher purpose.

I invite you to walk with me through this story, to see how resilience, faith, and love have carried me from humble beginnings to the incredible life I now live.

Lessons:

Family, faith, and integrity are the pillars that sustain us. The importance of love, service, and faith in guiding life's decisions is emphasized, reminding readers that personal values are the foundation of leadership and success.

3

Turning point: from healing to harvesting

'In seed time learn, in harvest teach, in winter enjoy' – William Blake

Everyone loves a story with sudden breakthroughs – the moment when everything shifts, when the seemingly impossible becomes possible. People often ask me, 'Benica, when and how did your life change?' Truth is, my turning point didn't happen overnight. It was a culmination of many small steps, failures, lessons, and a burning desire to forge a new path. But if I had to pinpoint that one defining moment, it would be in 2013, when I made a bold decision and finally ventured into farming.

Starting modestly in my backyard, I planted my very first crop – 20 cabbages.

It might sound trivial, but for me, it was a giant leap. I remember crawling like a baby, trying to stand tall on my new legs, eager to do something meaningful. I fell many times – literally and metaphorically; dusting myself off, learning from each mistake, and then trying again. Looking back, I am grateful I chose not to give up. Those early trial-and-error days, though tough, were the price I paid for the success I enjoy today.

The teething tears built the foundation of my knowledge and resilience. Many people also ask, 'Who mentored you?' or 'Who taught you farming?' Well, truth is, I didn't have a formal mentor in agriculture.

As Duncan once asked me during an interview, 'Who taught you farming?' I smiled and replied, 'Most of us Black people are natural-born farmers. It's something that runs through our veins, a skill passed down through generations, often learned through observation and inherited knowledge, rather than formal training. Africans have been producing their own food for centuries; storing the produce, drying vegetables for future consumption and preserving heirloom seeds for future generations.

In our community, farming isn't just a livelihood; it's a way of life. Our ancestors, the forebears of our land, farmed the same land that I now till. They enriched it with organic compost, nourished it with various kinds of organic manures and practiced what later came to be called crop-rotation. They had a deep relationship with the soil and the land. Though through my farming journey I did consult my grandmother from time to time, much of what I learnt was through my own curiosity and experimentation.

I remember countless hours spent exploring the land, feeling the texture of the soil, observing nature, and asking myself questions: How does this plant grow? Why do some crops flourish while others falter? The answers didn't always come quickly, but each experience added to my understanding of the symbiotic relationship between soil and crop.

Information technology has also transformed agriculture. Knowledge is not at the tip of our fingertips. There are various online resources such as search engines, social media and countless farming forums where one can exchange ideas with likeminded aspiring and aspirant farmers. It still requires experience to know what to do with information, when to act, and how to adapt. We live in an age of climate change and global warming; where climatic conditions of old no longer apply as they used to. This is where one requires to tap into other people's experience and guidance. It's at this point that's mentors and accountability come into play. I was fortunate to have elders and friends in the industry. Among those that quickly come to mind would be Augustine Pan Nemasisi, Sabzen Farmer, Matome Potatoe Farmer and many others who I look up to. They inspired me during my early days and I continue to look up to them.

I often say that my relentless curiosity is what gave me experience. I barely slept, spending hours researching online, reading everything I could find about farming, business, and entrepreneurship. I remember when Duncan and I were researching this book, I asked him, 'When do you find time to write?' He responded, 'I write at night'. We laughed, but it's true – my nights were often filled with late-night research, jotting down ideas, and dreaming about possibilities. I usually say in jest, 'Those who worship blankets seldom make it in this game,' because the extra hours and sleepless nights have been essential to my growth. Even in African culture there is a telling SiSwati idiom, 'Ingubo ayina mali' (A blanket will not give you money).

But success isn't just about working hard; it's also about working smart – having a plan, being strategic, and staying committed. I've learnt that one can be enriched with knowledge from observing successful people who, despite setbacks, keep their eyes on the prize. That's the essence of resilience. It's what pushed me forward when the going was tough.

One of my proudest moments was creating what would become the Benica Farm slogan: 'We farm, people eat'. It's a simple statement, but it captures the heart of my mission – providing food, hope, and opportunity to my community.

In the next chapter, we'll explore this slogan further and delve into the philosophy behind it.

Growing up as an orphan, I faced many hardships. Among those were limited resources, the absence of parental guidance, and a constant struggle to survive. Regardless, irrespective of those difficulties, I held on to hope – hope for a better future, for a life where I could make tangible difference.

My early years were marked by curiosity about land, life, and how to turn my circumstances around. That humble beginning ignited a passion for agriculture that grew into a movement, one that would inspire many others in my community to believe in the power of farming and entrepreneurship.

After spending eight years working in a hospital, I reached a point where I realized this wasn't my true calling. As a Department of Health employee, I was living the South African dream; stable job, steady income, a car, a house, secure retirement and a happy family. However, under that polished façade I was unhappy.

The toxic environment I worked in, self-hating and emotionally draining colleagues, bureaucratic frustrations, and a feeling that I wasn't fulfilling my purpose began to weigh heavy on me.

In our weekly meetings with Faith and Duncan, I often jotted down 'toxic people' in my journal. We'd laugh about it, but deep down I knew I couldn't keep living that way. One day, I told my co-writers, 'I'm leaving the hospital'. They looked at me in shock. They didn't understand why would I leave a guaranteed, stable career? Wouldn't I be risking everything? I explained that I was tired of the toxicity, of feeling trapped in a job that drained my energy and stifled my dreams.

My roots in farming had always been close to my heart, thanks to my grandmother, Nwa' Sikhemisi. She owned three pockets of land in our village, and I grew up watching her work tirelessly, tending her farms and providing for her family. I approached her for a piece of land, just a small section on one of her farms in Rooiboklaagte. She kindly demarcated a portion for me, trusting me to nurture it. That land became my testing ground.

At that time, I was already running small businesses – two spaza shops in Tsakani and Township; buying and reselling fruits, vegetables, and household essentials. I wanted multiple streams of income while I prepared to leave my job at the hospital. I was also planting a few cabbages in my backyard (remember the twenty I told you about?), experimenting with small-scale farming. When I finally decided to resign in 2012, I had my plan in place. I owned my spaza shops, had a modest piece of land given by my grandmother, and a payout cheque of R250,000. With all this, I felt ready to embrace my true calling.

Sharing my resignation story with Duncan and Faith, I realized the importance of having a vision that makes sense to you – even if it doesn't to others. My decision shocked many, including my boss, the CEO. I was confident because I knew I was making a move based on purpose, not just logic. At the time, it didn't make sense, but now it makes perfect sense. Statesman Nelson Mandela once said, 'It always seems impossible until it's done'.

That's the essence of entrepreneurship; things often don't make sense until they start making cents. My advice to aspiring farmers and entrepreneurs is, *Just start. The dots will connect as you go.*

As I transitioned from hospital worker to farmer and entrepreneur, the complexion of my days changed dramatically. No longer was I confined to the communications office, managing mundane tasks or resolving conflicts. Instead, I was in my element, building my businesses, tending my land, and exploring new opportunities.

My confidence grew as I started earning a salary equivalent to the one I was getting as a hospital employee; the important difference being that this time around I was doing something I love. Farming and entrepreneurship had become my new life.

The journey wasn't without setbacks. I faced doubts, financial hurdles, and the challenge of balancing multiple ventures. But each obstacle strengthened my resolve. I expanded my farm to ten hectares using a portion of my payout, bought my first bakkie to utilize in fetching water and moving supplies. I invested my payout in tools of my new trade and necessary infrastructure.

Another business I ventured into was buying and reselling cars; particularly old but reliable models. I bought my stock in Sabie and Hoedspruit. My first flip was an old Fiat Uno I purchased for R3 000.00, fixed it and resold for R15 000.00 in less than two weeks. That quick profit motivated me to continue exploring this avenue. I even bought the six-meter container I have already told you about to set up a fast-food business next to Dynamos Sports Ground in Tsakani.

We sold fish and chips. On weekends we hosted local DJs, turning the place into a lively gathering spot. That business taught me that with passion and perseverance, anything is possible. All these ventures – farming, car flipping, fast food, proved that I was capable of creating opportunities from nothing. I had the guts to show the world what I can do. With such great successes in other business ventures, it was easy for the idea of Benica Farm to be born.

It wasn't just a business; it was a symbol of my journey, a testament to resilience, faith, and hard work. The full story of how Benica Farm started, what it has become today, and the vision for its future will unfold in the next chapter.

But one thing remains clear: my journey from healing to harvesting was driven by a simple yet powerful desire – to build a life of purpose, impact, and abundance. And I believe that anyone willing to take that first step, no matter how small, can achieve greatness. All it takes is courage, a clear vision, and the willingness to start where you are.

What you gain:

The power of intuition and passion to pivot careers. It highlights that authentic fulfillment often involves listening to one's calling, risking comfort for purpose, and trusting that perseverance leads to harvests of success.

4

The birth of Benica Farm

'To forget how to dig the earth and tend the soil is to forget ourselves' – Mahatma Gandhi

It all began in the quiet corners of my backyard garden, where modesty was the only currency I had. A few small beds, some vegetables and a dream. That was early 2016 when I made the pivotal decision to fully commit to a life I had only begun to explore.

I clearly remember the day I took the leap: I was standing amidst the tiny sprouts of my garden, eyes fixed on the horizon, filled with a mixture of hope and uncertainty. I had already made up my mind that this was the new path I wanted to forge, one that would steer me away from the corporate world into the embrace of the land.

But no journey is complete without a fraction of doubt. I remember sitting with my co-writer, Duncan, during one of our many brainstorming sessions. He looked at me with a seriousness that caught me off-guard. He asked a question that would forever change my perspective, 'Why do you want to do this?'

'How do you see it working?' My response was honest and came from a space of confidence. My response put much emphasis on the 'why'. I knew that passion had already ignited a fire within me. When you are driven by passion, everything else – doubt, fear, obstacles – becomes secondary.

Doing what you love gives you a sense of pride, purpose, and fulfillment that no amount of money can buy.

That moment, when I answered Duncan, marked the beginning of a new chapter. I was determined to turn that backyard garden into something bigger, something meaningful.

I was willing to invest all my retirement savings to buy the necessary farming equipment and a trusty bakkie; and to start this new chapter with nothing but hope and a relentless desire to succeed. Even in those early days I could already see my future; lush green fields, abundant harvests, a thriving business that could uplift my family and community.

I still remember the first time I pictured myself as a full-fledged farmer – not just a hobbyist but a real agricultural entrepreneur. I could see myself walking through rows of cabbages, peppers, and tomatoes; feeling the soil beneath my fingers and knowing that every produce was a product of my hard work and perseverance.

That vision kept me going, and I wasn't going to let anything stop me. No doubts, no setbacks.

As weeks turned into months, I poured my energy into transforming the small backyard into a serious farm. I bought equipment, planting tools, irrigation systems, fertilizers, and started planting chili peppers, which had a high demand locally. The local spaza shops around the area began to place regular orders. I was receiving more and more requests daily, and I couldn't keep up alone.

The sense of achievement was intoxicating. Every morning, I woke up with a renewed sense of purpose; this was my new life, my new legacy.

But it wasn't just about planting and harvesting. It was about building something sustainable, something that could grow beyond what I imagined. I knew I needed help, and that's when I decided to involve my wife, Veronica in the business.

As I told you earlier, she was working at KFC, earning a modest salary. I saw potential in her and believed she could be the missing piece of this puzzle.

I had to convince her to leave her job, give up a steady paycheck and take a risk with me. I offered her a simple preposition; I'll pay her whatever she was earning at KFC, even if it was just peanuts. To my surprise, she agreed.

That decision marked a turning point. We opened our first small market stall next to the dusty street by Acornhoek Plaza. Our stall was adjacent to English-medium Wem Private School. It was a humble outlet; selling fruits and vegetables to locals who appreciated fresh produce. That first step was nerve-wracking but exciting. We duplicated the idea across the street near Mapulaneng TVET College. And before long, our little markets grew into something bigger.

We were no longer just farmers; we had become local vendors and entrepreneurs with a vision.

Our produce, fruits, vegetables, chilies and tomatoes became in high demand. We sourced some from our farm, others from small-scale farmers in Chochocho (Dingleydale) like Themba Makana, who became more than a supplier. He became a close friend and partner in this journey. Witnessing the birth of Benica Farm was more than just an achievement; it was a testament to community resilience, collaboration, and dreams turning into reality.

We pushed ourselves relentlessly. I remember selling at funerals, parties and community events. We went door-to-door, selling fresh produce to families who valued quality over price. Our commitment was fueled by hunger, not just for profit but for success, for recognition, for making a difference.

In that same year, Veronica and I officially launched Benica Farm. The name came from a combination of our names, Bennedicter and Veronica. I know today the name sounds like a marketing mindspark – but it was just that simple. It wasn't just a name; it was a symbol of hope, innovation, and community upliftment. The concept was simple yet powerful, a farm that will grow into a movement, a legacy to inspire others.

What started as planting twenty cabbages in a small patch quickly blossomed into a thriving farm. Today, just two weeks after transplanting, we harvest thousands of cabbages, peppers, tomatoes, spinach, and chilies. Each harvest a testament to perseverance, lessons learned, and continuous growth. Every leaf, every fruit, every vegetable is a symbol of what can happen when passion meets dedication.

Benica Farm isn't just about producing food. It's about showcasing sustainable practices, empowering local farmers and creating a culture of sharing knowledge. We wanted to do more than make a profit; we wanted to make an impact – transforming lives, creating opportunities, and inspiring others to believe in the power of agriculture.

As the farm expanded, I decided to reward Veronica for her unwavering support. I bought her a brand-new Toyota Avanza, a symbol of our progress. That moment; handing her the keys, was more than just a gift; it was a declaration that we were heading somewhere, that success was no longer just a dream but a tangible reality. I couldn't have afforded such a luxury during my corporate days, even after retirement.

Yet here I was, making that dream come true for my wife, proof that anything is possible.

But the journey didn't stop there. I had already started another business; providing school transport to learners with my trusty Nissan bakkie. I used that same bakkie for multiple purposes, transporting goods from the farm, moving our produce to markets and shuttling children to school. It became the backbone of my growing transportation enterprise.

Soon, my transport business expanded. I was transporting over 200 learners a week. My clients were pre-school, primary, and high school learners. It was intense; day in and day out. I soon realized I needed a bigger vehicle, so I invested in a Toyota minibus to better accommodate the growing number of learners. I also employed two drivers to help manage transport logistics. That allowed me to focus on expanding my other business ventures.

Demand for my services started to grow. I bought a second-hand KIA truck to transport larger quantities of fresh produce.

With all these efforts, I started supplying fresh produce to seven major retailers such as Tintswalo Super Spar, Thulamahashe Boxer, Acornhoek Rootz and others.

My social media presence played a crucial role. The more I shared my journey, the more people believed in me. I became sort of an influencer, a farmer with a story that resonated far beyond my local community.

A turning point came in 2018 when Duwyne Du Plessis of Acornhoek Mall SuperSpar, discovered me. Duncan, in our book, mentioned that Du Plessis saw me on the road near his residence and recognized my potential.

When he was preparing to open a new Super Spar at Acornhoek Mall he reached out to me. He saw the quality of my produce and believed I could be a reliable supplier to his supermarket. That was the beginning of a new chapter – one that would test my reliability and commitment.

He offered me the opportunity to supply their supermarket; a chance I didn't take lightly. I started small, with watermelons, but over time the demand grew exponentially. We faced growing pains, overwhelmed by orders, struggling to keep up with quality, delivery delays, and resource constraints. Du Plessis was honest with me, highlighting my shortcomings and advising me to focus on fewer clients, on quality, and consistency.

It was a tough lesson; realizing that growth sometimes demands sacrifice, focus, and discipline. I had to let go of trying to serve everyone and instead dedicate my energy to becoming a trusted main supplier. That advice changed everything. I understood that to truly succeed, I had to prioritize quality over quantity, consistency over haste.

Our relationship with Du Plessis grew stronger. We built trust through reliability and quality. I learnt to maintain standards, even during challenging times. There were moments of tension, deductions from payments for late deliveries or subpar products, but those setbacks only strengthened my resolve. I realized that maintaining high standards is essential in business, especially when working with big brands.

Throughout this journey, I was also supporting small-scale farmers; buying from them, circulating money within our local community. Our farm became more than just a business; it became a hub of activity, cooperation, and growth. We believed in lifting each other up; creating a ripple effect of empowerment.

But with success came challenges. Once again, jealousy, sabotage and even betrayal started to seep in. soon there were toxic individuals who saw my progress as a threat. I remember a time when I was supplying Du Plessis, and some staff and community members tried to sabotage my efforts. They falsely claimed I lacked proper transportation documentation, attempted to disrupt my services and spread rumours.

It was painful seeing how people who should support me instead chose to undermine my progress.

One of the most hurtful incidents involved an insider; a staff member who forged an invoice worth R22 000 using my identity.

He made arrangements with my driver, someone I trusted, to make deliveries. After a normal workday, the driver, under false pretense, delivered to a client products I didn't supply, claiming they were from Benica Farm. The sabotage scam was uncovered through CCTV footage, revealing a conspiracy to steal my hard-earned money.

The culprits were caught and faced justice, but the damage lingered. It made me realize that sometimes your greatest enemies are people closest to you – fellow Africans, fellow Black men and women, who, instead of supporting each other, choose to tear each other down. I've pondered this deeply; is poverty the root cause? Or is there a deeper underlying issue: our own internal struggles with unity, pride, and progress? I don't have all the answers, but I know that Black-on-Black sabotage is a barrier we must collectively overcome.

Despite these setbacks, I refused to give up. I learned valuable lessons about resilience, integrity, and the importance of community. I knew that my journey with Benica Farm was bigger than just me – it's about creating a legacy, a movement for empowerment and growth. We've crossed many hurdles, but each obstacle only made us stronger.

Today, Benica Farm has evolved beyond just planting crops. We are now into agro-processing – producing jam, chili sauces, herbs, and other value-added products. We've established systems and processes that allow us to turn raw produce into packaged goods, expanding our reach and impact. We are no longer just farmers; we are entrepreneurs, innovators, and community builders.

What you've read up to now is just a glimpse of what Benica Farm has become – a symbol of hope, resilience, and collective effort. It's the story of a young couple with a shared vision: to build something meaningful, something that will outlive us and inspire future generations.

From humble beginnings to national recognition, from local markets to international partnerships, our journey is far from over.

This chapter, like the farm itself, is a testament to dreams realized through hard work, perseverance, and unwavering belief. It's about turning setbacks into lessons, failures into stepping stones, and passion into legacy. The future holds even greater promise, and I am excited to see where this journey continues.

Takeaways:

Starting small with passion can grow into a sustainable enterprise. This chapter teaches that dedication, community collaboration, and strategic planning are essential for turning dreams into reality.

5

None Genetically Modified Food and sustainable agriculture

'Small farmers feed the world' – La Via Campesina

The Role of Water and Nutrients in Crop Production

1. The Lifeline of Plants: Water and Nutrients

For plants, water and nutrients are not optional – they are essential for growth, health, and productivity. Roots absorb both water and dissolved nutrients from the soil. As plants draw water, it's naturally replenished by rainfall or manually through irrigation.

In natural systems like forests and grasslands, nutrients cycle back into the soil through decaying plant litter; fallen leaves, stems, and other organic matter. This self-sustaining loop ensures soil fertility. However, on agricultural land, we harvest crops and remove most of the plant material; grains, leaves, roots – which means we also remove valuable nutrients. Without replenishment, the soil becomes depleted, and crop yields drop.

To maintain soil fertility and ensure healthy crop growth, nutrients must be replaced. This is where fertilization – either chemical or organic, comes in.

2. Fertilizers: Choosing Between Chemical and Organic Options

In South Africa, most commercial farmers rely on chemical fertilizers, purchased from agricultural suppliers. These products are nutrient-dense and easy to apply, especially with mechanized equipment.

However, for many small-scale or rural farmers, chemical fertilizers pose challenges:

They're expensive.

Access is limited, especially in remote areas.

Transporting them is labor-intensive.

Despite these barriers, nutrient replenishment is a non-negotiable. Poor soil = poor crops, no matter how much effort you put in.

Kraal Manure: The Traditional Alternative

Before chemical fertilizers, farmers around the world, including here at home, used manure to maintain soil fertility. Kraal manure is organic material made up of digested plant matter from animals kept in kraals.

Manure is less concentrated than chemical fertilizer but can be highly effective, especially on smaller plots. It's a natural, accessible, and affordable solution. For small-scale farming, kraal manure often provides enough nutrients [NPK] to sustain healthy crop growth.

This nutrient can be applied either through broadcasting, drenching or foliar applications

Tip: *Make Your Own Liquid Fertilizer from Kraal Manure*

Instructions:

Soak 50 kg of kraal manure in 200 litres of water in a tightly sealed container.

Let it ferment for 7 days to allow beneficial microorganisms to break down the organic matter.

After 7 days, open the container and dilute the concentrated mixture at a ratio of 5 litres per 20 litres of clean water.

Apply the diluted liquid directly to the base of your plants as a soil drench.

Result: This nutrient-rich liquid fertilizer will significantly boost plant health and growth. Watch your crops thrive like magic. Once you try this method, you'll never buy fertilizer again

3. What Nutrients Do Plants Really Need?

Three key nutrients drive plant development: NPK

Nitrogen (N): Essential for leaf development, chlorophyll production, and overall plant growth.

Phosphorus (P): Plays a vital role in root development, flowering, and fruiting.

Potassium (K): Contributes to overall plant health, disease resistance, and the development of strong stems and stalks.

4. Soil Testing: An Investment Worth Making

Soil testing allows you to identify nutrient deficiencies and tailor your fertilization plan. Samples sent to a lab are analysed, and experts recommend the exact type and amount of fertilizer needed for your specific crop.

While soil tests cost less than R500 per sample, the long-term benefits; higher yields and efficient fertilizer use are worth the investment.

Skipping soil testing is a risky gamble. Investing upfront helps you avoid wasting time, money, and effort on crops that underperform.

Pro tip: *Test your soil before planting. It's smarter to prevent nutrient problems than to fix them mid-season. Treating Farming as a business not a hobby.*

5. Restoring Depleted Soils: The Manure Cure

If your land has been overused without proper fertilization, it's likely nutrient-starved. Signs include:

Poor growth

Small, pale, or yellowish leaves

Thin stems

Low yields

The solution? A manure broadcast.

Spread manure evenly across your field. For effective restoration:

Apply 250 wheelbarrow loads per hectare

Or 1 wheelbarrow per 40 m^2

This will help rebuild the soil's nutrient base and improve future yields.

6. Maintaining Fertility: Fertilize Before You Plant

Once your soil is revitalized, keep it fertile by replacing the nutrients your crops remove. Apply kraal manure before planting.

Different crops remove nutrients at different rates. High-yielding crops like maize or potatoes extract more nutrients than crops like spinach or legumes.

Always fertilize according to the crop's nutrient needs.

7. Dealing with Acid Soils: Lime and Ash to the Rescue

In high-rainfall regions, soils often become acidic, making nutrients unavailable to plants; even when manure is added. This results in disappointing yields.

To fix this, apply agricultural lime:

Use 2 to 3 tons per hectare to neutralize soil acidity.

If lime is unaffordable or hard to access, use wood ash from cooking fires.

Mix ash with your kraal manure before application.

Ash raises soil pH and helps make nutrients available for uptake.

Soil pH matters! Consider testing for acidity before investing in fertilizers.

8. Don't Let Weeds Steal Your Nutrients

Weeds compete with your crops for water and nutrients. In fertile soils; especially those enriched with manure, weeds thrive just like your crops.

If you don't control weeds:

They will consume most of the added nutrients.

Your crop yields will suffer.

You'll end up growing a lush garden of weeds instead of food.

Key takeaway: *Fertilize + weed control = success. Neglect either, and you'll likely be disappointed.*

9. Companion Planting and Crop Rotation: Farming Smart

Crop rotation and companion planting:

Reduces disease and pest pressure.

Helps manage soil fertility naturally.

Improves yields over time.

Invest time in learning these techniques; they are powerful tools for sustainable farming.

10. Watch Out for Pests and Diseases

Even with perfect soil and water, crops can still fall to pests and diseases. Control requires a multi-pronged approach:

Monitor your field regularly.

Identify problems early.

Use integrated pest management (IPM) strategies.

We will provide detailed pest control guidelines in upcoming chapters tailored to commonly grown crops by small-scale farmers.

Kindly note: *Soil is a living resource. Treat it with care. Feed it, protect it, and understand it; and it will feed you in return.*

Always remember: Good farming begins with healthy soil. Let's consume healthy food, None Genetically Modified Foods, caring for the environment and sustainable agriculture should be every farmer's concern.

Spraying and Fertilizer program for Cabbage production on 1 Hectar for commercial purposes

MONTH	PRODUCT	QUANTITY/DOSAGE	HOW IT IS DONE	FUNCTIONS
	MONTH BEFORE PLANTING		DO SOIL PREPARATION	
3 days before planting	Make ridges Apply lime/gypsum Apply NPK 4:3:4	2 metres between the rows 500kg/ha 2000kg/ha	Tractor ridger or physical Apply on the soil Broadcasting or placement	Neutralize Ph Mix the fertilizer with soil and irrigate
week at planting	Seedlings Kick start 4 Cypermenthrin Chloropirifos wetcit	28 000-35 000/ha 400ml+ 30ml+ 40ml+ 10ml in 200l water	60cm in between plants Drench plants	Root development Cutworms Soil borne pest Sticker and wetter
Week 2	LAN(28) Nitrospray plus Calmabon plus Coppercount Steward wetcit	400kg/ha 400ml+ 400ml+ 400ml+ 10ml+ 10ml in 200l water	20g/plant Spray plants	Application of nitrogen Foliar feeding Worms Sticker and wetter
Week 3	Delthamethrin Chloropirifos Agtap Steward Abamectin wetcit	30ml+ 40ml+ 12g+ 10ml+ 12ml+ 10ml in 200l water	Spray plants	Ballworm Worms Insects Worms Leafminer Sticker and wetter
Week 4	LAN(28) Nitrospray plus Calmabon plus Coppercount Steward Chlorothanonil(mycoguard) wetcit	400kg/ha 400ml+ 400ml+ 400ml+ 10ml+ 50ml+ 10ml in 200l water	20g/plant Spray plants	Application of nitrogen Foliar feeding Worms Fungus diseases Sticker and wetter
Week 5	Imidacloprid(confidor)	20ml+ 40ml+		Whitefly and aphids

	Chloropirifos Agtap Steward Abamectin congfu wetcit	12g+ 10ml+ 12ml+ 30ml+ 10ml in 200l water	Spray plants	Worms Black diamond moth Worms Leafmi ner Bacteri al diseases Sticker and wetter
Week 6	LAN(28) Nitrospray plus Calmabon plus Coppercount Steward Chlorothanonil(myc oguard) wetcit	400kg/ha 400ml+ 400ml+ 400ml+ 10ml+ 50ml+ 10ml in 200l water	20g/plant Spray plants	Applica tion of nitrogen Foliar feeding Worms Fungus diseases Sticker and wetter
Week 7	Imidacloprid(confid or) delthamethrin Chloropirifos Agtap Steward Abamectin congfu wetcit	20ml+ 30ml+ 40ml+ 12g+ 10ml+ 12ml+ 30ml+ 10ml in 200l water	Spray plants	Whitefl y and aphids ballwor m Worms Black diamond moth Worms Leafmi ner Bacteri al diseases Sticker and wetter
Week 8	NPK 1:0:1 Nitrospray plus Calmabon plus Coppercount Steward Chlorothanonil(myc oguard) wetcit	400kg/ha 400ml+ 400ml+ 400ml+ 10ml+ 50ml+ 10ml in 200l water	22g/plant Spray plants	Head development Foliar feeding Worms Fungus diseases Sticker and wetter
Week 9	NPK 1:0:1 Imidacloprid(confid or) delthamethrin Chloropirifos Agtap Steward Abamectin congfu wetcit	400kg/ha 20ml+ 30ml+ 40ml+ 12g+ 10ml+ 12ml+ 30ml+ 10ml in 200l water	22g/plant Spray plants	Whitefl y and aphids ballwor m Worms Black diamond moth Worms Leafmi ner Bacteri al diseases Sticker and wetter
Week 10	NPK 1:0:1 Nitrospray plus Calmabon plus Coppercount	400kg/ha 400ml+ 400ml+ 400ml+	22g/plant	Head development Foliar feeding

	Steward Chlorothanonil(myc oguard) wetcit	10ml+ 50ml+ 10ml in 200l water	Spray plants	Worms Fungus diseases Sticker and wetter
Week 11	NPK 1:0:1 Imidacloprid(confid or) delthamethrin Chloropirifos Agtap Steward Abamectin congfu wetcit	400kg/ha 20ml+ 30ml+ 40ml+ 12g+ 10ml+ 12ml+ 30ml+ 10ml in 200l water	22g/plant Spray plants	Whitefl y and aphids ballwor m Worms Black diamond moth Worms Leafmi ner Bacteri al diseases Sticker and wetter
Week 12	NPK 1:0:1 Nitrospray plus Calmabon plus Coppercount Steward Chlorothanonil(myc oguard) wetcit	400kg/ha 400ml+ 400ml+ 400ml+ 10ml+ 50ml+ 10ml in 200l water	22g/plant Spray plants	Head development Foliar feeding Worms Fungus diseases Sticker and wetter

NB: Adhere to safe use and handling of agrochemicals. This program is subjected to soil analysis and change of weather conditions. Strictly follow label!

Always know that cabbage belongs to heavy feeder family, it requires sufficient nutrients and water to give you outstanding results. If you can use overhead irrigation, cabbages love overhead irrigation.

What to learn:

The importance of soil health, organic practices, and environmental stewardship. It emphasizes that sustainable farming is vital for food security and long-term community wellbeing, urging farmers to prioritize ecological balance.

6

Leadership, community building & mentorship

'In seed time learn, in harvest teach, in winter enjoy' – William Blake

This chapter naturally flows into the next, but it also stands as a testament to a truth I hold dear: leadership isn't just something you learn, it's something you are born with. From my younger days I found myself stepping into roles that demanded strength, compassion and resilience. As the saying goes, a leader is not made but born. It is a sentiment echoed by many who have observed or interacted with me over the years. Interviews with respected voices like Themba Makana, Masingita Zwane, David Byrne, Duwyne Du Plessis and many others consistently highlight the leadership qualities I embody.

It's humbling to hear these reflections, but I believe they're rooted in genuine experiences. My leadership journey began long before I understood what it truly meant. Do you remember the football team I mentioned in the first chapter? That was more than just a game; it was my first real opportunity to lead. I was fortunate to captain a group of talented young players; boys who looked up to me not just because I was the team captain, but because I inspired them.

I didn't realize at the time that I was demonstrating qualities that would shape my future; qualities like organization, dedication, teamwork, and responsibility. Football was more than a sport; it was a lifeline, especially for a kid from the rural areas where access to activities and resources was limited. It was a window to camaraderie, discipline, and hope.

The bond I formed with my teammates has endured over the years. To this day, I maintain close relationships with some of those friends, and that's no coincidence. My experience on the field taught me invaluable lessons about building communities, understanding other people, and leading with humility. It was in those early days that I learnt that leadership is about service, putting others before yourself and fostering an environment where everyone can thrive.

My journey in leadership continued into high school, though the details are a bit hazy and I can't recall the exact year. But I still remember being elected a class prefect; the feeling, the sense that I was chosen among many because I embodied something special. I was to be elected multiple times, each time leading by example, always striving to do my best.

These early experiences reinforced a crucial lesson: when called to lead, questioning why is secondary. The question that should fill your mind is, how do I serve? How do I make a difference? I wore the shoes of leadership not because they fitted perfectly, but because no one else was stepping into them. And I believed that I could make a difference.

In those formative years, I led with passion and purpose. I was driven by a desire to uplift others, to bring people together, and to create a sense of unity. Many people who have interacted with me since then still say the same thing, I am someone who instills trust, inspires confidence, and builds bonds. I've disappointed some along the way – I'm human, after all; but I have also gained the trust of many. Winning hearts is no small feat; it's an honor that I cherish deeply. Despite challenges, I remain relevant and hopeful, especially in the eyes of the young ones who see me as a beacon of possibility.

Some even place bets on me, believing I can lead them into a better future.

Reflecting on my mother's passing and my father's absence, I realize that leadership often manifests in moments when we least expect it. When I was left to care for my four younger siblings, I was thrust into a role I hadn't prepared for; to be their protector, provider, and guide. That was leadership in its rawest form. It was not a choice but a necessity.

I had to step up, even when I felt scared or overwhelmed, because someone had to. That experience shaped my character, instilling a deep sense of responsibility and trustworthiness.

My faith community also played a significant role in shaping my leadership style.

Growing up in the IPHC, fellowshipping at its Greenvalley branch, introduced me to leadership quite early through church activities. From a young age, I found myself circulating in leadership roles - from youth league positions to representing the church at various gatherings.

I didn't fully understand at first what it meant to lead; I simply followed the call. As I matured, I became more conscious of my influence, making deliberate choices to lead with integrity.

When I was appointed to lead the newly opened Cottondale branch, I was still young but felt a strong sense of purpose. Moving my family there was a significant step, but I embraced it wholeheartedly. The influence I had in that community was undeniable. People saw in me a leader who genuinely cared. There's a popular saying that 'A leader without followers is just a man taking a walk'. I was leading a new congregation, I had followers, a congregation I had to guide and uplift.

I was elected to represent the youth league at the provincial level, a role that further reinforced my belief that leadership is about service, empathy, and collaboration.

Duncan, my writing collaborator, once asked me why I was chosen for those leadership roles, especially in Cottondale where my record wasn't marked by extraordinary achievements.

I believe it's because my presence, my energy, and my authenticity spoke volumes even before I said a word. People recognize genuine leadership - the kind that radiates positivity and trust. I've always carried a positive aura, a vibe that influences those around me. When I step into a room, I can feel the energy shift; people are drawn to that light, eager to follow it.

Leadership, I've learned, extends beyond church halls and community gatherings. It's woven into every aspect of life. In my professional journey at the hospital, I was entrusted with leadership roles that challenged and refined me.

As chairperson of the hospital office administration, overseeing all clerk in the hospital, I had to demonstrate resilience, strategic thinking, and compassion. Though young, my impact was clear —

sometimes it was met with praise, sometimes with criticism. But every experience strengthened my resolve and helped me understand the multifaceted nature of leadership.

My impact isn't limited to formal roles. I am a natural giver, and that trait defines much of my community work. Over the years, I've quietly contributed to charitable causes, feeding families, supporting those in need, not for recognition but because I understand what it feels like to have little. I remember the days when I had only one meal a day, when I was vulnerable and unsure. That empathy fuels my desire to give back. During the festive season, my wife and I try to give generously, spreading joy and hope.

It's not about showing off; it's about making a difference, however small, in someone's life.

My grandmother, Nwa' Sikhemisi, played a pivotal role in shaping my attitude towards giving. She taught me that true generosity comes from the heart. Giving without expectation of return, especially when others lack, opens the door to blessings - an idea I've come to trust deeply.

I believe in the universal law of exchange: what you give out, whether love, kindness, or resources, will find its way back to you. This belief guides my actions and informs my approach to community service.

Growing my love for planting and growing fruits and vegetables was born from that same spirit.

It's a form of giving, a way to nurture life and foster sustainability.

In 2025, my wife and I founded the Benica

To: Mr B Mhlongo

The Department of Agricultural Economics and Agribusiness at the University of Venda is deeply grateful to Benica the Farmer. Thank you for your insightful and engaging presentation on the Entrepreneurship and Career Development Webinar on 08 May 2025.

This experience has significantly enriched colleagues' and students' understanding of entrepreneurship within the agricultural sector.

We truly appreciate you sharing your expertise and knowledge with our audience. Your presentation was well-received, followed by much positive feedback. We hope to explore opportunities for future collaboration.

From: Department of Agricultural Economics and Agribusiness

University of Venda

Foundation, a charity dedicated to empowering, supporting, and strengthening community resilience. This initiative is a movement rooted in empathy, collaboration, and shared purpose.

My leadership style continues to evolve, but at its core, it's anchored in empathy and collaboration. I mentor emerging farmers, university students, and community members, sharing practical skills, sustainable practices, and entrepreneurial insights.

Through programs at Benica Farm, I aim to cultivate leadership, resilience, and hope. I believe that community-driven initiatives have the power to transform lives — creating a ripple effect of positivity that can uplift entire communities.

For me, leadership is about problem-solving. When people come to you with questions and challenges, they're testing your ability to find solutions. Sometimes, they're testing your limits or your intelligence. That's why a leader must be sharp, continuously seeking growth and knowledge.

I spend late nights reading, learning new skills, and staying informed about politics, economics, agriculture; anything that keeps my mind active and ready to face new challenges. The world changes daily, and so should we. You should ask yourself whether you are evolving as well? And whether you are becoming better every day?

Understanding and valuing people is fundamental to good leadership. People don't care how much you know until they know how much you care. It corresponds with that spiritual maxim that at the end of our days God will not ask us which car we drove, but how driven we were. Touching hearts and building genuine relationships is the foundation of influence. Leaders don't just share their agenda; they listen, understand, and respect those around them. That's how trust is built, and trust is the currency of effective leadership.

In conclusion, leadership isn't just about holding titles or making speeches. It's about embodying integrity, compassion, resilience, and humility. It's about building communities, mentoring others, and leaving a legacy of hope and positive change. Lest they say, 'Often a leader is a man who knows the road but cannot drive the car'. It's called leading from behind.

As I reflect on my journey, I am reminded that every step I've taken was driven by a deep sense of purpose - to serve, to uplift, and to lead by example. Because at the end of the day, true leadership is about making a difference in the lives of others often quietly but profoundly.

Lessons:

Leadership is rooted in service, humility, and genuine care. Building strong communities and mentoring others creates a ripple effect of growth and resilience. It advocates that true leaders cultivate trust and foster collective empowerment.

7

Collaborations, recognition & awards

'Support your local farmer. You can't eat oil' – Unknown

Throughout my journey in the farming industry, I have been fortunate to cultivate a strong network of support, partnerships, and friendships that have shaped the success of Benica Farm. From brands and businesspeople to everyday farmers and community leaders, I have built bridges across many walks of life. My social media platforms; Facebook, TikTok, X, LinkedIn et al have become my digital farm, where I share my story, insights, and tips with thousands of followers beyond borders. What started as a simple backyard garden, planting just 20 cabbages has exploded into a thriving operation producing over 15,000 cabbages in just two weeks.

This growth is a testament to one vital truth: success is born from nothing more than grit, innovation, and resilience. I didn't inherit wealth, I didn't receive special favors, nor did I have rich investors backing me from the start.

What I had was an unshakeable belief in my vision and the willingness to work tirelessly, learning as I went. It's a story of humble beginnings that many can relate to, an ordinary person with an extraordinary dream.

As my social media presence grew, a door opened; an invitation for collaborations, partnerships, and mentorships with various brands and institutions came through.

My pages became a platform not just for storytelling but also for genuine engagement with communities, students, and entrepreneurs.

I remember collaborating with Bokamoso Academy, an accredited institution committed to education.

The goal was to provide short courses for their learners, which would also serve as a practical extension of what we do at Benica Farm.

Building on this, Semila, a fertilizer company specializing in agricultural inputs, came into the picture, offering services and support that enriched our short courses. These relationships are rooted in genuine trust and shared objectives.

Today I have formal collaborative agreements with institutions of higher learning such as Ehlanzeni TVET College, specifically the Mapulaneng campus, the University of Mpumalanga, the University of Venda, Tshwane University of Technology, Mangosuthu TVET, and even international universities.

Through these partnerships, I train students and empower emerging farmers with skills and knowledge that can change their lives. It's about planting seeds of hope – literally and figuratively, in the minds of future generations.

Recognition, both locally and internationally, has followed my efforts. Many people wonder what motivates me to share every moment of my life on social media; whether it's work, personal milestones, or everyday struggles.

Some ask why I don't keep my life private. My answer is simple, I believe in transparency, in sharing my journey authentically, because it inspires others. It's not about fame or recognition for its own sake; it's about making an impact.

When I walk into a room and people shake my hand or bow, it's not because of my influence alone but because of what I stand for – the tangible impact of my work.

People tell me that they check my social media profiles daily, not just out of curiosity but because they find inspiration in my journey. Young farmers, small-scale farmers, and aspiring entrepreneurs see me as proof that it's possible to rise from nothing and build something meaningful.

I remember posting a simple question; 'Are you ready for this book?' and receiving an avalanche of positive responses. Many couldn't wait to get their hands on it. If you are one of those supporters, I want to express my heartfelt gratitude. Your encouragement, your words of support, your messages have fueled my determination to see this project through.

My story has been long overdue. I recall an inspiring moment during an interview with Duncan, when Themba Makana said, 'Bennedicter, they pushed you to tell your story because if you don't, who will?' That resonated deeply. I knew then that I had to make it happen, not just for myself but for countless others who look up to me.

Wednesday 6 July 2022

www.dailysun.co.za

13

FROM NURSE TO FARMER!

Ben wants to help youth

SUCCESS

By OHIS MHASI

BENNEDICTER Mhlongo is a nurse by profession, but he gave that up 10 years ago to be a farmer.

Today, the 36-year-old runs BenicaFarm in Acornhoek near Bushbuckridge, Mpumalanga, on 10ha of farmland that creates jobs for 15 people.

"Our main crops are tomatoes, green pepper, chillies, cabbage and spinach. We supply our produce to leading retailers such as SuperSpars, Lebamba in Hoedspruit, school feeding schemes and hawkers at shopping centres."

He said the farm was recently certified compliant by SA Good Agricultural Practices (GAP) audits.

"We have a generation that is clueless about producing its own food, which is wrong. Young people don't want to hear anything about agriculture because of the stigma that it is for uneducated and old people. We need to change that."

Bennedicter said with help from the government, they could assist young people to realise that farming could be done by the youth.

"Farming is expensive and without assistance, you won't make it. To create enthusiasm, government should consider investing in youth in farming," he said.

"Such opportunities should benefit people who are in farming and not comrades and their families or friends."

He said his farm provided practical experience to people who were studying at institutions of higher learning.

"Farming is in my blood. What I know I'm always willing to share. As someone who grew up without parents and raised by a gogo, I exist to prove to the world that a black child can run a global GAP-certified farm," he said.

"I exist to prove to the world a black child can run a global GAP-certified farm"

Bennedicter Mhlongo is proud and happy that his farm has created job opportunities for a few people (inset).

I wanted to answer their questions, to share my journey in a way that leaves no stone unturned. I hope this story doesn't leave you hanging, but instead, invites you into the depths of my experiences, my struggles, my triumphs, and my relentless pursuit of a better life.

Recognition from national agencies has also come my way. The National Youth Development Agency (NYDA) in Mpumalanga discovered me through social media and reached out. Mr. Mnisi, a dedicated manager, saw potential in my work and invited me to a meeting at their Mbombela headquarters.

From that day, a partnership was born; one that has transformed my farm into a hub of mentorship for their beneficiaries under the Benica Farm banner. In the past year, we have mentored young farmers, helping them develop skills, access resources, and find their footing in agriculture.

This partnership was a turning point - an acknowledgment that my work was making a real difference. It wasn't just about growing crops; it was about growing communities. Since 2024, we have been working closely with NYDA, creating opportunities for youth and small-scale farmers to thrive.

Beyond NYDA, numerous brands and organizations have supported us; Indalo Inclusive South Africa, Timbavati Foundation, SEDA, Fetola, and more. Their support has been instrumental in acquiring essential resources, tractors, solar panels and irrigation systems that powered up our farm.

I continue to attend workshops, collaborate on projects, and build relationships with these partners because I believe in the power of collective effort. These partnerships are not just about funding; they are about shared visions and mutual growth.

My co-writer Faith Ngwenya was tasked with interviewing Farai Mauchaza, the Finance Coordinator and Nedbank Indalo Fund Manager, about our relationship with Indalo.

Farai recounted how we first met in 2021 after I responded to a Call For Applications for the Sustainable Tourism Value Chains Enterprise Support Program, implemented by Indalo in partnership with Old Mutual and Mont Dor. That initial grant of R15000.00 was just the beginning.

I received business coaching, technical support, and ongoing mentorship, which helped me expand and refine my operations.

Representing Indalo, Farai expressed pride in my work. Their support over four years has brought positive change not only to Benica Farm but to the entire community. I've created jobs, increased food security, and demonstrated that small-scale farming can become a sustainable business.

When opportunities arise, I am among the first to seize them - not because of favoritism but because I meet the criteria shaped by these organizations. They see in me a reflection of what they hope to nurture: resilience, innovation, and social impact.

Thanks to this solid relationship, I have also attracted funding from major players like SAB, Nedbank, SEDA, Fetola, and others. These organizations see the potential in Benica Farm - not just as a profitable venture but as a platform for empowering small-scale farmers and transforming communities. Each investment is a vote of confidence in our mission to create sustainable, impactful change.

Let me seize this opportunity to also share the story of David Byrne and his wife, Arda Van Dongen. They moved to South Africa seven years ago, seeking new adventures and a fresh start. Byrne, a retired Canadian, and Arda, a Dutch national, fell in love with South Africa's warmth and hospitality.

During an interview, Byrne shared how South Africa became more than just a vacation destination; it became their new home. Their love for the country and its people was evident in their ongoing support for initiatives like ours.

He also highlighted what many of us believe: the importance of supporting those who are pushing for positive change in their communities. Byrne discovered me through social media in 2021 and decided to contribute, first with a water pressure pump to help with irrigation, then with ongoing support and encouragement. Their contributions have gone beyond mere resources; they've helped build confidence and exposure for Benica Farm.

I cherish the words of encouragement from people like Byrne. They remind me that my work resonates far beyond the farm. I recall the day Byrne gently reminded me about an upcoming interview, his way of showing ongoing support. That small gesture meant a lot. It's these human connections, these acts of kindness, that breathe life into my journey and my story.

Social media, as I've mentioned before, has been a powerful tool. It's where I tell my story, where I share proof of my progress. I believe in storytelling with evidence; showing the farm, sharing milestones, and being transparent about struggles and successes. That's how I've built a large following and gained recognition from brands and individuals who believe in my vision.

When I announced that we were writing this book, I didn't expect the overwhelming positive feedback. I wasn't trying to seek validation but rather preparing my supporters for this long-awaited project. My approach is strategic; I understand that storytelling is not just about sharing but about inspiring action.

Beyond my farm, I've been part of a movement that's close to my heart; the Bushbuckridge Farmers Association (BBRFA). This initiative was born from a long history of collective effort among farmers in the region. It was pioneered by the late Frank Fumani Mnisi, a passionate farmer, entrepreneur, and advocate for Black empowerment. The movement started as a simple WhatsApp group; a forum for farmers to share ideas, pool resources, and support each other. They ran a 'stokvel', a communal savings scheme to buy livestock and farming materials, aiming to grow together.

Over time, interest waned, and the movement lost momentum. But after Frank's passing, I was called upon by his family to deliver a eulogy at his funeral. It was a moment of reflection and unity.

Many people thought I and Frank were at odds; our disagreements in business and farming had created that impression. But beneath those differences was a deep friendship, rooted in shared goals and mutual respect.

After the funeral, I proposed that we formalize the movement into an organization. That's what later became the Bushbuckridge Farmers Association. With proper governance, a constitution, and regular meetings, we transformed it into a robust movement with over 200 active members. We now conduct site visits, training sessions, and collaborate with government departments and Non-Governmental Organisations.

We recently hosted a community fun run – a 5 and 10 km marathon starting from Orpen Gate, sponsored by Wits Rural Facility, Timbavati Foundation, Acornhoek Mall SuperSpar and local businesses.

These events foster unity, promote healthy lifestyles, and raise awareness about agriculture and rural development. I believe that when farmers and communities come together, they can overcome challenges like climate change, market access, and scarcity of resources.

My efforts have not gone unnoticed. I've received several awards that motivate me to continue.

These include the Social Entrepreneur Award from FNB & Fetola in 2023, recognizing my innovative approach to farming and community development; the Climate Change Adaptation Award from Indalo Inclusive South Africa, acknowledging my sustainable practices; and recognition from the Mpumalanga Department of Agriculture for regional contributions.

While I appreciate these accolades, I want to emphasize that my motivation isn't fame or awards. It's about feeding people, creating opportunities, and leaving a legacy of hope and resilience.

For me, success is measured in lives changed, communities uplifted, and futures secured.

My journey continues; focused not on popularity but on impact. I want to see small-scale farmers thrive, food security increase, and young entrepreneurs find their footing. That's my true reward: knowing that my work makes a difference.

I aim to leave behind a legacy that whispers hope, courage, and kindness to future generations – something more enduring than Likes or Followers on Facebook.

This chapter is just a glimpse into a story still being written. The real journey is only beginning.

What readers take:

Success is amplified through partnerships and authentic engagement. The chapter celebrates humility and the importance of giving back, showing that recognition often follows dedication and community-focused efforts.

8

Overcoming challenges & embracing growth

'Farming is a profession of hope' – Brian Brett

Every journey worth taking is riddled with obstacles; resource shortages, unpredictable weather patterns, limited access to markets, societal perceptions, and sometimes even outright hostility. When I reflect on my path, I realize that these hurdles are not just random setbacks but essential parts of the story that shape resilience and character. It's easy to get discouraged when faced with such difficulties, but I've learned that resilience and a growth mindset are the keys to navigating these turbulent waters.

For me, setbacks are not failures; they are opportunities - opportunities to learn, adapt, and innovate. Each challenge has taught me something valuable. Whether it's a drought that tests my water management skills or market rejection that forces me to diversify, I see these moments as stepping stones rather than stumbling blocks.

Continuous learning; seeking new knowledge, skills, and perspectives has kept me afloat. Community support has been a lifeline, and faith in my vision has kept my spirit alive.

My story, I believe, is an invitation to every aspiring farmer out there; see challenges not as dead ends but as stepping stones.

Never underestimate the power of perseverance. It is through persistence, through facing the storm head-on, that we grow stronger and more confident.

Now, let me take you a little further back into my journey. As my farm and small business ventures grew, so did the presence of toxic influences – people whose intentions were not always pure.

When I first started the fruits and vegetables business, I was filled with hope and excitement. I had opened a small vendor stall, then another in Boelang, a nearby village. Business was booming, and I was making good sales. I felt a sense of pride in what I was building.

But as the business gained momentum, trouble arose. Locals in Boelang started complaining.

They accused me of killing their small businesses, claiming that my success was somehow threatening their livelihood. These were local traders and small-scale vendors who relied on street selling and roadside stalls to make ends meet.

They felt squeezed out, and their grievances quickly turned into accusations.

What made it more complicated was that a political party, Better Residents Association (BRA), claimed to stand for the rights of the community and small businesses. They argued that my business was unfair competition and that I was destroying the local economy. I was caught in the middle of a storm I didn't fully understand.

Despite my intentions to create opportunities and grow my business, I was forced to shut down my Boelang stall. I had invested heavily in building that shop, money, time, and hope. And seeing it come to an abrupt end was painful.

But life teaches us resilience in many ways. I took that loss as a lesson, one that reinforced the importance of understanding local dynamics and community sensitivities. I decided to return to Rooiboklaagte, where I grew up and where I had my roots.

It was a tough decision, filled with frustration and disappointment, but I knew that the journey was far from over.

Moreover, I still experienced the very same hate when I opened another stall in Acornhoek, opposite Acornhoek Library along the main road. The landlord started developing hate when he saw my business booming. He started making unfounded allegations to my employees. He later decided to evict me, saying I must remove my business which was inclusive of a building I have invested a lot of money to build. I then decided to move out and left the shop.

In retrospect, I often wonder; why should I be responsible for the success or failure of someone else's business? Most of us have access to similar resources, and yet, some blame others for their setbacks.

It's a common tendency to point fingers rather than take responsibility. I believe that success or failure ultimately depends on individual effort, innovation, and perseverance.

Sometimes, I felt that people saw me as a threat; more than just a competitor. They saw me as a symbol of change, progress, and perhaps, their own limitations. In some cases, I sensed that jealousy and fear of losing control fueled their hostility.

Sadly, this hate extended beyond local rivalry and small traders. It grew into a broader resentment, often fueled by racial or economic tensions.

Some local people began blaming foreign nationals – Indians, Somalis, and other entrepreneurs who owned small shops or operated businesses in their communities.

They accused these foreign nationals of destroying local ventures. But the truth is, no one has stolen from you; no one has taken your business by force. The real issue is that we often fail ourselves. We blame external factors or others for our failures instead of asking, 'What can I do differently?'

Enough about those conspiracy theories. The real lesson is that in business and in life, setbacks are inevitable. What matters most is how we respond to them. I've learned that resilience is built not in comfort but in adversity.

Another significant challenge I faced was collaborating with private investors during the COVID-19 pandemic. I remember reaching out for support to fund a large-scale project – planting 50,000 cabbages on a budget of over R200,000. It was during a time when hope was thin, and the world was fighting an invisible enemy.

Our plan was ambitious, and I was eager to make it work. I recruited investors who preferred to remain anonymous, trusting that together, we could create something meaningful.

However, climate change proved to be our greatest adversary. Just two weeks before harvest, a torrential downpour, something I hadn't seen in years destroyed all our cabbages. The heavy rain flooded the fields, washing away our hard work and leaving me in debt. We had signed a clear contract with the investors, with agreed-upon percentages and responsibilities. I had promised to be accountable for the outcomes, but I wasn't prepared for Mother Nature's wrath.

This was a first for me. I had never experienced extreme weather conditions that wiped out an entire crop. I couldn't recall a storm of that magnitude in my farming years, not since the year 2000. It was heartbreaking. I remember sitting in my small office, staring at the muddy fields, feeling helpless and overwhelmed. If I had to write an essay about a day I will never forget, that day would be at the top of the list. It was a day filled with regret, frustration, and despair.

The financial repercussions were severe. I had to start paying back the investors, and the burden weighed heavily on me. My bank started repossessing some of my assets—including my small KIA truck since I couldn't cover the installments or insurance. During the strict lockdown, things got even tougher. The government announced that all repayments for bank-financed vehicles and homes could be paused; a small mercy, but not enough to erase mounting debts.

I fought hard to keep my possessions. My Toyota Avanza was almost taken, but I borrowed money from close friends to pay off the remaining R21,000 and held on to my vehicle. Those months were some of the most difficult in my life. I felt defeated, broken even. But I learned a lot from those losses; lessons that no amount of success could teach me.

Some lessons came painfully; lessons about financial management, the importance of diversified income streams, and the resilience needed to survive tough times. I wouldn't wish such hardships on my worst enemy, but I also understand that adversity is a part of growth. Sometimes, setbacks are necessary for us to appreciate the value of perseverance.

Let me share a story that underscores challenges faced by farmers, from someone I deeply respect, Pollen Mokgope. I first met Pollen online, and we quickly bonded over our shared passion for agriculture. He's from Welverdiend, a place much like my own farm. Pollen specializes in poultry and livestock. He is a successful farmer, the kind of person who has weathered many storms and still stands tall.

I asked Pollen if he could share his insights on farming during challenging times. He agreed, and I tasked Faith to interview him. Pollen's words resonated deeply with me. He said, 'Being a farmer is tough. Sometimes it feels like a job with no reward. But despite the hardships, it gives you purpose. Watching your animals grow healthy and free from disease gives a sense of fulfillment that's hard to describe. There's a strange boredom that settles in after you sell your products – you feel like you're not doing enough. That feeling motivates you to work harder, to increase your production, even when it's tough'.

Pollen's perspective reminded me that farming is more than just a business; it's a calling. It's about patience, resilience, and a deep love for the land and animals. Challenges may come, but they also bring opportunities to learn, improve, and serve a higher purpose.

As I reflect on my journey, I realize that every obstacle has shaped me. From dealing with community hostility, weather disasters, financial setbacks, to personal doubts – each experience has strengthened my resolve. I want to share these stories with you because I believe that failure is not the end. It's a stepping stone. And growth often comes after the hardest battles.

To every farmer, entrepreneur, or dreamer reading this; know that your journey is unique. Embrace challenges, learn from them, and keep moving forward. Resilience, faith, and a willingness to adapt are your greatest tools. The road may be rough, but the rewards of perseverance are priceless.

Key lessons:

Resilience transforms setbacks into opportunities. Challenges like resource scarcity, sabotage, and climate disasters teach adaptability, patience, and the value of perseverance. It encourages viewing obstacles as lessons on the road to growth.

9

The gospel of farming

'Farming is not just a job, it's a way of life' - Anonymous

For me, farming isn't just about planting seeds in the soil or harvesting crops; it's a divine calling – an act that has the power to transform lives, uplift communities, and create a legacy that endures beyond a single lifetime. When I look at the land, I see a canvas of endless possibilities, a place where dreams can take root and flourish. Farming, in its purest form, is an act of hope – a testament to resilience, patience, and unwavering faith in the future. Growing up in rural communities, I witnessed firsthand how farming serves as the backbone of our society.

It's not merely a livelihood; it's a way of life that binds people together, instilling a sense of purpose and belonging. It's the heartbeat of the community, pumping life into every corner of the land, every home, and every family that depends on its bounty. Over the years, I've come to see farming as more than an economic activity; it's a sacred mission to nourish the world and nurture the human spirit. My mentorship programs are born from this belief.

I dedicate my time and energy to empowering youth and emerging farmers, because I understand that the future of our communities depends on the seeds we plant today. Planting seeds is both literal and metaphorical. We plant the seeds of knowledge on young people to see agriculture as a way of life. And we hope those seeds grow and they embrace production of food, not consumption as a role they should need to play for the rest of their lives. Because if there's one thing that will never be altered by the passage of time; it's that people will always need food – and for food to be available there's always be the need for a farmer. That's why the joke holds that 'You cannot insult a farmer with food in your mouth'.

These empowerment programs focus on more than just teaching the technical aspects of farming; they aim to ignite a passion, foster entrepreneurship, and develop strong leadership. When young people learn sustainable practices – like conservation farming, crop rotation, and water harvesting, they gain confidence. When they see their efforts translate into meals on their tables and income in their pockets, their hope is renewed.

Through storytelling, I share my own journey – the struggles, the setbacks, and the triumphs – to show that farming is not a shortcut to easy riches but a pathway of perseverance. I encourage farmers to see failure not as a defeat but as lessons that strengthen their resolve. Practical training sessions are designed to equip them with skills needed to adapt to changing climates and market demands. We share experiences around the fire, exchange ideas, and celebrate each small victory along the way.

Farming also teaches us about patience; waiting for the right season, nurturing the soil, and trusting the process. It's a humbling but rewarding journey that reminds us that growth takes time. That patience, coupled with innovation and resilience, is what separates a successful farmer from one who gives up at the first obstacle.

But beyond the personal growth of individual farmers, I believe agriculture must be the vehicle that drives national development. Our land has the potential to feed not only our families but also entire communities, regions, and even countries.

The demand for food is unending – an eternal cycle that sustains life itself. Imagine a society where every unemployed individual is actively engaged in farming activities. What if we harnessed this potential, transforming idle hands into productive, purpose-driven lives?

In such a society, crime and despair would diminish because people would find fulfillment through meaningful work.

Instead of wandering aimlessly or resorting to negative influences, they would be cultivating the land, growing their own food, and contributing to local markets.

This sense of ownership and purpose creates a ripple effect - families become more self-sufficient, communities thrive, and the economy gains momentum.

I envision a future where urban and rural areas work hand in hand; cities supporting local farmers, and farmers supplying fresh produce to urban dwellers. Urban farming initiatives, community gardens, and cooperative markets could become commonplace, bringing food closer to the people and reducing dependency on imported food. A country without a food security strategy rooted in local production is a country doomed to fail.

This approach not only ensures food security but also nurtures a culture of sustainability and respect for the land.

Farming is also a powerful tool for social cohesion. When communities come together to work on shared projects; planting communal fields, building irrigation systems, or organizing harvest festivals, they strengthen bonds and foster a sense of unity. These activities become opportunities for mentorship, skill transfer, and cultural preservation.

They remind us that despite our individual struggles, we are part of a larger collective with a common purpose: to build resilient, prosperous communities.

Moreover, the potential of farming extends beyond food. It can be an avenue for innovation – integrating technology such as mobile apps for market prices, drone surveillance for crop health, and solar-powered irrigation systems. These advancements make farming more efficient and accessible, especially for young farmers eager to embrace modern solutions. By embracing innovation, we can bridge the gap between traditional knowledge and contemporary practices, creating a sustainable future for generations to come.

However, I often hear people say, 'Farming is hard, it's risky, and it's uncertain'. Yes, it demands hard work, patience, and resilience. But what worthwhile endeavor doesn't? The rewards – nourishing your family, earning an income, and contributing to the well-being of others are worth every sacrifice. Farming teaches us humility and gratitude, reminding us that life's greatest treasures are often found in the simplest of things: a handful of healthy soil, a sprouting seed, a ripe fruit.

I believe that if we start viewing farming not just as a job but as a sacred gift, we can inspire a new generation of farmers who see their work as a noble calling. We must change the narrative - celebrating farmers as heroes, innovators, and custodians of the land. Their work is vital; it sustains the very fabric of society.

In conclusion, farming is more than a livelihood; it's a gospel - a message of hope, resilience, and renewal. Let us embrace it with pride, passion, and purpose. Let us nurture the land and our communities, knowing that through farming, we plant the seeds of a brighter, more sustainable future. Together, we can turn the soil of adversity into a fertile ground for growth, prosperity, and everlasting hope.

What to learn:

Farming is a divine calling—a mission of hope and nourishment. It fosters community, teaches patience, and is a catalyst for national development. The chapter inspires viewing agriculture as a sacred, impactful profession that sustains the human spirit.

10

The future of farming

'Small farms are the backbone of a sustainable food system and the heart of rural communities' – Unknown

The future of farming in Bushbuckridge and beyond isn't just a distant dream; it's a vibrant, unfolding reality – full of innovation, sustainability, and inclusiveness. As I stand here today, reflecting on my journey, I see a landscape brimming with possibilities. I envision a future where mentorship blossoms, climate-smart agriculture becomes the norm, and a new generation of farmers rises to carry the torch of growth and resilience. My life, in many ways, feels like a movie – one filled with unexpected twists, hard lessons, and moments of breakthrough that have shaped who I am.

From my early days working as a nurse, quietly tending to other people's health, to now leading communities and tending to my farm, each chapter has imparted lessons about adaptability and purpose.

I've learned that life doesn't always follow a straight line; sometimes, you have to rewrite the script to find your true calling.

My ultimate dream is simple yet profound: to harvest breakfast and lunch right from my own land. To see my farm flourish with crops that nourish my family and community. But beyond my personal goals, I believe that the future of farming holds the promise of creating the next generation of billionaires – people who will transform agriculture into a thriving, innovative industry that sustains and uplifts entire communities. Agriculture in Bushbuckridge needs to become the economy it is and contribute to the GDP of the municipality in a substantial way.

One idea I hold close to my heart is establishing a distribution hub right here in Bushbuckridge – a central place where small-scale farmers and young producers can bring their harvests, connect with markets, and ensure their hard work doesn't go to waste.

This isn't just about business; it's about building a system where money circulates among us, strengthening our local economy and empowering our people.

I often think about how difficult it can be to find a simple ingredient like chili when I need it.

If I want to buy or resell chili, I often struggle to locate a reliable source. It's frustrating, sometimes products spoil before they reach the right hands because of lack of access to proper markets or distribution channels.

That's why I believe a distribution hub would change everything. It would be a game-changer; making it easier for farmers to sell their produce, for consumers to access fresh goods, and for everyone involved to thrive.

Imagine a space where any farmer, whether just starting out or already seasoned, can bring their harvest and be confident it will reach the right market.

No more products rotting in the fields due to lack of buyers. No more days wasted searching for buyers or worrying about spoilage.

Instead, farmers can focus on what they do best – growing; while the hub takes care of distribution, seamlessly connecting supply with demand. This idea didn't come out of nowhere. It started from my own experiences at Benica Farm, where I saw firsthand how a well-organized system could transform the way we farm and sell.

I envisioned a future where our farm isn't just a place of production but a hub of activity – where ideas and resources flow freely, and opportunities abound.

Fortunately, I've begun to see this vision come to life. I've been fortunate enough to connect with private investors from Germany who understand my passion and share my vision.

We're in the early stages now, working tirelessly to turn this dream into reality. It's a complex process, but I am confident that with perseverance and collaboration, we will succeed.

But this isn't just about infrastructure or markets; it's about empowering farmers.

Many small-scale farmers are afraid to produce certain crops because they don't know where to sell them or how to find buyers. They have the skills and knowledge to grow these products, but the absence of a reliable market discourages them. They fear investing their time and resources into crops that might not bring a return.

This is where the distribution hub becomes a lifeline. It will serve as a bridge – connecting farmers directly with buyers, markets, and even exporters. It will create opportunities for farmers to diversify their produce, experiment with new crops, and grow with confidence. It will also help in building trust and relationships within our community – showing that farming isn't just about producing food but also building sustainable livelihoods.

Looking ahead, I see a future where young people in Bushbuckridge see farming not just as a means of survival but as a pathway to prosperity. I want to inspire them to see the agricultural sector as a place of innovation and opportunity.

With the right support, mentorship programs, and access to resources, the next generation of farmers can revolutionize our landscape – adopting new techniques, embracing technology, and making farming a lucrative and respected profession.

I often reflect on my own journey – from a young orphan in Rooiboklaagte to a thriving farmer and community leader. It's a story of resilience, hope, and the unyielding belief that change is possible.

I want others to see that their circumstances don't define their future; that with determination and community support, they can rise above challenges and achieve greatness. As I look to the horizon, I am filled with optimism. I see a future where farming is sustainable, inclusive, and driven by innovation.

I imagine community markets bustling with activity, young entrepreneurs experimenting with new crops, and families enjoying nutritious meals made from fresh, locally-grown produce. I envision a landscape transformed not just by technology or infrastructure but by the strength of our community and our shared dreams.

Farming is more than just planting seeds; it's about cultivating hope, nurturing communities, and growing a better future for all of us. I believe that together, we can build a brighter tomorrow, one where the land continues to give, and our people receive and thrive.

So, as I stand here today, I remain committed to this vision.

The journey is ongoing, and there will be challenges ahead, but with faith, perseverance, and the support of our community, I know that the future of farming in Bushbuckridge will be extraordinary. It's a future worth working for – full of promise, potential, and the power to transform lives.

If government is truly committed to addressing the triple challenges of poverty, unemployment, and food insecurity, then agriculture must be seen as a key solution. Food is a daily necessity-consumed three times or more a day, which means the demand for it will never fade.

I therefore urge young people to seriously consider agriculture as a viable career path or source of income. Together we can tackle food insecurity, reduce unemployment, and uplift communities out of poverty.

Farmers, business leaders, government, and all relevant stakeholders, let us join hands and work together to reshape the small-scale farming industry.

'The future millionaires and billionaires of Africa will come from agriculture' – Dr Akinwumi Adesina, President, African Development Bank

That's my story. The journey continues.

Takeaways:

Innovation, inclusiveness, and sustainability define the future. It emphasizes the importance of mentorship, technology, and local market development, inspiring readers to see farming as a lucrative, impactful industry that can transform societies.

Acknowledgments

Let me take this opportunity to express my deepest gratitude to all those who have walked beside me on this journey - those who believed in the vision long before it bore fruit.

To my family; thank you for understanding when I'm constantly on the move, day and night, chasing a dream and building this empire. Trust me, every step I take is for your benefit. One day, all the hard work will pay off.

To my wife, your love, patience, and unwavering encouragement gave me the strength to keep going, even when the path felt uncertain. You are my anchor.

To my mentors; Mr. Rest Kanju of Indalo Inclusive SA. Ms. Candice Pierce of Timbavati Foundation and Mr. Ntando Maseko of Fetola - I am a successful entrepreneur today because of the three of you. Your guidance, wisdom, and belief in me helped me unlock my purpose. I am forever indebted to you. To the communities who welcomed me, listened to my ideas, and helped turn dreams into reality - especially those in the Hoedspruit area, some of you deserve special mention: Mr. David Byrne, Sarah Dawn Bergs, Richard van Schalkwijk, Stuart Fairbairns, Derick Du Doit, Jason & Angelia, Danny Versloot & his wife, Marianne - each of you played a critical role in my growth. To Acornhoek SuperSpar Manager Duwyne; your trust and support for our products as local farmers gives me courage to excel in everything I'm building. I will forever be grateful. To Ms. M.E. Zwane, former CEO of Tintswalo Hospital; Dr. M.P. Thuketana, former Clinical Manager; Ms. Virginia Mzimba, former Hospital Board member; Mrs. N.J. Dlamini and Ms. M.B. Moyimane, the then Hospital Nurse Manager – and the entire hospital management team. You helped sharpen a strong spear in me. Thank you for understanding my calling when I asked to be released to pursue another path. Your support gave me the courage to take a leap of faith.

To those who simply heard about me and chose to reach out, offering help, opening doors, or sharing a kind word; your belief has made all the difference. To my grandmother, N'wa S'khemisi, who raised my siblings and me from childhood into adulthood after we lost both our parents, I will forever be humbled by her strength and selflessness. I take off my hat to her. To all our employees; the team behind every success story, without you, there is no Benica Farm. Thank you for your hard work and dedication. Let us continue serving with pride and purpose.

This book, this journey, and the gardens that continue to grow are the result of collective effort – rooted in a shared vision and nurtured by community spirit.

To all my social media followers

Much respect!

About the author

Bennedicter Mhlongo is an award-winning farmer, first generation SA-GAP (Good Agricultural Practices) certified farmer, social entrepreneur, mentor, and respected leader. Born in 1984 in the dusty streets of Roiboklaagte, Acornhoek, his journey from humble beginnings to national recognition is both inspiring and transformative. His passion for agriculture began in 2012 with a simple backyard garden. But long before he traded scrubs for soil, Bennedicter had an established career in healthcare. After matriculating in 2002, he trained and worked as a security officer, before enrolling in Sekororo Nursing School in 2003. Upon graduating, he served as a nurse at the same hospital for two years before moving to Tintswalo Hospital. There, he held diverse roles„from working in male surgical, maternity, and TB wards to serving in the Hospital Matron 's office, communications office and the hospital's CEO. Though his dedication to healthcare was evident, Bennedicter's heart was always in farming. In 2013, he made the bold decision to resign from nursing and fully pursue his agricultural dreams„a journey he describes as moving ,from nursing patients to nursing plants.' Today, Bennedicter is the proud founder of Benica Farms, a thriving agroprocessing enterprise that began as a backyard project and now supplies produce to major retailers and markets. But his impact doesn't stop at farming. 103 He is a committed advocate for small-scale farmers, serving as the Chairperson and one of the founding members of the Bushbuckridge Farmers Association NPC, where he champions rural development and agricultural entrepreneurship. He also serves as an Advisory Board to Mpumalanga Youth In Agriculture and Rural Development[YARD] He actively mentors agri-students from institutions such as: " " " " Tshwane University of Technology (TUT) University of Mpumalanga (UMP) TVET Colleges NYDA crop beneficiaries A devoted husband to Veronica Mhlongo, Bennedicter is not only a farmer but a family man, leader, and inspiration to many. His story is a powerful testament that with passion, perseverance, and purpose, anyone can rise above their circumstances. He lives by a simple yet unwavering rule: ‚From childhood to adulthood, only one rule: Never give up.'

Testimonials

We have produced solid future generation of farmers, a magnitude of Alumnus. In the days of my life, I have touched people's lives in a positive and sustainable way.

I owe a great deal of my farming journey to Benica, whose guidance and generosity changed the course of my career. When I was just starting out, unsure and overwhelmed, he took the time to teach me the basics not just of farming techniques, but of patience, resilience, and respect for the land.

What stood out most about Benica was his willingness to share knowledge without hesitation. Whether it was walking me through planting cycles, helping me troubleshoot soil issues, or simply offering encouragement on hard days, he was always there with wisdom and a steady hand. 13:05

< 251 Trevor

Mr Benica i will like to take this moment to thank you for what you have done in my life it was not any easy journey some otherday i was feeling like you were putting us under pressure but you were teaching us that you have to work hard in order to achieve something at the end of the day and being consistent in your work can gives you better results at the end . here i am a graduate today all thanks to you🙏 15:56

To: Mr B Mhlongo

The Department of Agricultural Economics and Agribusiness at the University of Venda is deeply grateful to Benica the Farmer. Thank you for your insightful and engaging presentation on the Entrepreneurship and Career Development Webinar on 08 May 2025.

This experience has significantly enriched colleagues' and students' understanding of entrepreneurship within the agricultural sector.

We truly appreciate you sharing your expertise and knowledge with our audience. Your presentation was well-received, followed by much positive feedback. We hope to explore opportunities for future collaboration.

From: Department of Agricultural Economics and Agribusiness

Letata Farm 22 Jul
to Bennedicterm

Warm greetings dear Patriarch of farming, among many epistles and testaments of farming in South Africa , You have caught our eyes as Youth Farm Talk Lesotho committee to request your honourable presence on the 26October 2024 in Lesotho, Botha Bothe District just a few kilometers after crossing Caledonspoort bridge.

All expenses will be covered for you, we just request you to prepare a speech so as Basotho Youth grow in Farming. We propose you arrive on the 25th October ,accomodation will be covered as well.Hoping to hear from you soon.

www.ingramcontent.com/pod-product-compliance
Lightning Source LLC
LaVergne TN
LVHW090419210926
848480LV00002B/621
* 9 7 9 8 2 3 1 2 8 5 0 5 1 *